University of Pennsylvania Annual Report of the Provost to the Board of Trustees

From September 1st, 1899
to September 1st, 1900

Elibron Classics
www.elibron.com

Elibron Classics series.

ISBN 0-543-91470-4 (paperback)
ISBN 0-543-91469-0 (hardcover)

This Elibron Classics Replica Edition is an unabridged facsimile
of the edition published in 1900 by University of Pennsylvania, Philadelphia.

THE UNIVERSITY BULLETINS NEW SERIES—NO. 2.

UNIVERSITY OF PENNSYLVANIA

THE PROVOST'S REPORT

FOR THE YEAR ENDING

AUGUST 31, 1900

Founded 1740

PHILADELPHIA
PUBLISHED BY THE UNIVERSITY
DECEMBER, 1900
BI-MONTHLY

UNIVERSITY OF PENNSYLVANIA

Annual Report

OF

HE PROVOST

TO THE

BOARD OF TRUSTEES

From September 1st, 1899, to September 1st, 1900

PHILADELPHIA
PRINTED FOR THE UNIVERSITY
1900

To the Trustees of the University of Pennsylvania:

GENTLEMEN:—I have again the honor of presenting to you an account of the work and fortunes of the University during an eventful and busy year. The year ending August 31, 1900, covered by this report, was marked by the happy completion of important undertakings, which had their inception in former years, by the beginning of new movements tending towards the high University ideal which we have set before us, and by many occurrences of more than passing interest. As is inevitable, we have had personal losses to mourn, and it is my first duty to make this reverent and affectionate record of those whom death has taken from us.

My last report told of the satisfaction with which we regarded the acquisition of Dr. J. M. Da Costa as a member of the Board of Trustees: this must tell how brief the term of service permitted to him. Brief as it was, it was wise, loyal, and effective from beginning to end. To it Dr. Da Costa brought a clear and well-trained mind, a ripe experience, a profound and practical knowledge not only of his profession, but of the requisites and methods for its successful teaching; and above all that conscientious and broad acceptance of his Trusteeship, which embraced all University interests, and made its affairs a part of the business of his life. It will not be forgotten that he and a friend laid the foundation of a fund, which, when sufficiently increased, will enable the University to grant retiring allowances to members of the Faculty who shall have grown old in faithful service.

Dr. John Ashhurst, Jr., was forced by failing health to surrender the chair of Surgery and of Clinical Surgery, which he had filled with distinguished honor, and accepted from you the Emeritus title, after thirty-three years of connection with the Medical School. Not long afterwards he was called away, leaving behind him a memory and an influence that will not be soon forgotten. That Dr. Ashhurst was a surgeon of high ability goes without saying. Thoroughly independent and conservative, he was not always in accord with the theories and methods of his brother surgeons; but the measure of success that attended his practice, the accepted authority of his writings, and the testimony of able surgeons who proclaim themselves disciples of Ashhurst, are a sufficient testimony to his wisdom and skill as a practitioner and a teacher. Dr. Ashhurst was a man of deep and sincere religious convictions: I do not think that these were obtruded upon his colleagues and students, but I think that these were always aware of them, and that whether consciously or not, patients and pupils and colleagues trusted and revered the more the man whose life was governed by them.

Full of years and of honors, another Emeritus Professor was taken from us by the death of Dr. Alfred Stillé, sometime Professor of the Theory and Practice of Medicine, and of Clinical Medicine. The name is classic in our University. In my last report it was my duty to speak of what Dr. Charles J. Stillé was to it as the accomplished scholar, learned in literature and history, and as Provost. Not less cultured in the humanities, his brother, Alfred Stillé, won his reputation by his splendid medical work, as investigator, author and teacher. Only your special request induced him to retain the chair of Medicine for several years after he had decided to resign it, and to make way for a younger man.

When he finally retired from teaching, he gave his large and valuable Medical library to the University, with the desire that it might still be helpful to the work and service to which he had given twenty years of his life. That library is now a memorial of one who had no small share in maintaining and advancing the fame of our Medical School.

Still another Emeritus title ceased with the death of Professor James Parsons. For twenty-four years Professor Parsons held the chair of Commercial Law and Contracts in the Law Department, and was noted in the Faculty as one whose studies leaned rather to the fundamental and original principles of Law. So earnest was he in the desire to lead his students into these studies that he induced many to take them up in what was practically seminary work, to which he was willing to devote any amount of his time and care. In the course of time this special work was recognized, not altogether wisely, as a course by itself, and for some years rewarded by the degree of Master of Laws. By recent concerted action between the Faculties of Law and of Philosophy, the post-graduate study of Law, for which Professor Parsons contended, is properly provided for; and the degree of Master of Laws will be given under conditions similar to those which govern other post-graduate degrees.

In my last report I had occasion to refer to the murder of an instructor in our Law School, and it is with feelings that I cannot express that I record a tragedy of the same kind, but perhaps more horrible. Roy Wilson White was one of our graduates, whose undergraduate history marked him as one likely to reflect credit on the school as a member of its teaching force. Upon his work he entered with high hope and enthusiasm. At the end of a long day's work, late in the evening, on his way from the Law School building to his home, he was set upon by three men, and robbed and murdered.

Dr. Samuel P. Molenaer, whose death occurred early in the summer, was for little more than a year instructor in French in the College. In that brief service he won the regard of his colleagues, not only by his fidelity to his teaching duties, but by his readiness to go beyond what was perfunctory, and to give his best efforts to whatever would inspire interest in his department or promote that fellowship which is so essential to effective work.

The reconstruction of the Surgical teaching on the same lines as that of the preceding year for Medicine led to a similar series of appointments in that subject. Dr. J. William White, the Professor of Clinical Surgery, was the logical successor to Professor Ashhurst, and you have therefore elected him to the "John Rhea Barton Chair of Surgery." Two Clinical Chairs were created, and to them were elected Dr. Edward Martin and Dr. Charles H. Frazier, both of whom were trained under surgeons who have held office in recent years. Dr. Martin and Dr. Frazier had gained, in subordinate positions, the practical experience which has warranted their promotion to the responsibilities of Clinical Chairs under the new system.

There are now three of our teaching staff on leave of absence for the purpose of important public service. Assistant Professor Emory R. Johnson has been for a year past a member of the Isthmian Canal Commission, and was selected by the President to study and report on the prospective economic results of the opening of such a canal. When the question of the good government of our recently acquired territory came up, the President appointed Assistant Professor L. S. Rowe one of the Commissioners to prepare a code of laws for Porto Rico, and Professor M. G. Brumbaugh Commissioner of Education for that island. While, of course, these appointments present gratifying testimony to the esteem in which members of our Faculty are held, and while the University is glad to lend its aid in the solution of the important problems with which they are dealing, much care has to be given to providing for their work in such way that our students may suffer no loss.

STATISTICS OF TEACHERS AND STUDENTS.

The number of Professors, Lecturers and Instructors is two greater than that recorded one year ago. The total is 260. There has been a decrease of 117 in the number of students—almost entirely in the Department of Medicine. The increased entrance requirements have had the expected effect of reducing, for the time being, the number of successful applicants for admission to this Department.

The statistics classified, as in my last report, and compared with those of the preceding year, are as follows:

PROFESSORS, LECTURERS AND INSTRUCTORS.

	1898–99.	1899–00.
The College	106	104
Department of Philosophy	55	54
Department of Law	14	14
Department of Medicine	91	94
Laboratory of Hygiene	3	3
Department of Dentistry	39	41
Department of Veterinary Medicine	18	18
Emeritus Professors	5	5
	331	333
Duplications	73	73
Total	258	260

STUDENTS.

	1898-99.	1899-00.
The College	926	968
Department of Philosophy	158	172
Department of Law	320	312
Department of Medicine	835	682
Laboratory of Hygiene	19	19
Department of Dentistry	502	484
Department of Veterinary Medicine	50	46
	2,810	2,683
Duplications	20	10
Total	2,790	2,673

There were registered at the University, during the year under review, 92 students from foreign countries, a decrease of 12; 815 from States and Territories other than the State of Pennsylvania, an increase of 76. From the State of Pennsylvania itself we received 1,858 of our student body, a decrease of 89.

I subjoin, as usual, a table giving the geographical distribution of our students for the year 1899-1900:

Geographical Summary of Students.

Africa	1	Italy	1	Pennsylvania	1858
Alabama	2	Ireland	1	Peru	1
Argentine Republic	1	Japan	2	Porto Rico	1
Arizona	2	Kansas	6	Portugal	1
Australia	9	Kentucky	10	Rhode Island	15
California	17	Leeward Islands	1	Russia	3
Canada	20	Louisiana	3	Scotland	1
China	1	Maine	21	South Carolina	5
Colombia	2	Maryland	20	Spain	1
Colorado	3	Massachusetts	56	Sweden	1
Connecticut	27	Mexico	3	Switzerland	3
Costa Rica	4	Michigan	5	Tennessee	6
Cuba	3	Minnesota	2	Texas	11
Delaware	37	Mississippi	5	Utah	3
Dist. of Columbia	5	Missouri	12	Vermont	4
Ecuador	2	Nebraska	5	Virginia	6
England	5	New Hampshire	10	Washington	4
Florida	1	New Jersey	170	West Indies	2
France	4	New Mexico	2	West Virginia	10
Georgia	4	New York	112	Wisconsin	12
Germany	8	New Zealand	7	Wyoming	1
Hawaii	1	Nicaragua	4		
Holland	1	North Carolina	11		2683
Illinois	21	North Dakota	1	Duplications	10
Indiana	11	Ohio	59		
Iowa	1	Oregon	2	Total	2673

THE COLLEGE.

The Dean begins his report with a feeling reference to the resignation of Professor Barker. The many years Professor Barker has spent in the service of the College, and the importance of his subject, warrant both the extent and the tone of the Dean's notice. I shall not dwell here upon this, but shall content myself with referring to the passage in the Dean's report, as well as to the similar passage in that of the Dean of the Department of Philosophy.

The illness of Professor Barker, as well as that, happily not so prolonged, of Professor Laird, necessarily threw increased work upon the assistants in their departments. And it is with peculiar pleasure that, following the example of the Dean, I note for your special attention the zeal and efficiency with which these assistants took up and bore forward the work when their departments were thus short-handed. It was upon the shoulders of Assistant Professor Goodspeed and of Mr. Pilcher that the responsibility, and consequently the chief burden, in these two departments fell. It is gratifying to know that the abilities and success of Mr. Pilcher in his difficult task have met with the recognition they deserve in an advanced position in our historic institution.

The most important change that has been made during the past year is probably the readjustment of those courses which are more particularly intended to prepare college students for the study of Medicine. The proper adjustment of the College course, at least in the upper years, to the courses of the professional schools is a problem that has been often approached of late years, though as yet no universally satisfactory solution of it has been reached. It is partly, not wholly, a question of time. It has been felt that if, as seems desirable, the student who enters upon a course of professional study should have first graduated from the College, there must necessarily result, with things as they at present are, an unjustifiable expenditure of time. He comes to the actual work of his profession too late: his apprenticeship has trespassed upon what should be the term of his independent life. But this is really only a partial view. Time is important, and should not be wasted. Waste in the expenditure of effort is perhaps more important and more serious. From this point of view the question becomes this: May it not be possible, without sacrificing anything that is valuable educationally in the College course, so to modify it that it may be the helpful forerunner of the professional course that is to follow? And if this can be guarded, will not the result be a clear gain, in so far as continuity is secured, for the student from the beginning of his College course to the close of his professional training? This secured, the man who has taken a College course so planned, and followed it up with a course in the professional school, would unquestionably find himself, not only in a general sense better educated, but with a fuller and better professional equipment than his fellow who had skipped in his haste the College course. These were the ideas that led to the arrangement in the College a year or two ago of courses in the Junior and Senior years to meet the wants of such as desired ultimately to take up the study of Law. With a similar purpose, other courses had been laid down as preparatory to Medicine. These were the Chemistry-Biology group in the courses in Arts and Science, and the two-year special course in Biology. During the past year the attention of both the College and Medical Faculties was called again to these courses and to the relation between them and the first year of the course in Medicine. It did not take long to see that, with the development that Chemical and Biological studies had reached in the College, we had in them not only in general a preparation for the Medical course, but within certain limits an actual equivalent for portions of it. It was also recognized by the College that as the first year of the Medical course is at present arranged, educational as well as professional character attached to it. And so it has come about, happily with results that can hardly fail to justify the action, that the College now recognizes educational value in studies of the first year of the course in the Medical School, and the Medical School recognizes professional value in certain studies of the upper years in the College. It is possible, then, according to this adjustment, for a student, under certain provisions, to be a Senior in the College and a Freshman in the Medical School simultaneously. There has been no sacrifice of principle, only a recognition of patent facts. So far as Medicine is concerned, by looking for a reduction of waste in effort, we have secured also a saving in time. For the details of the plan, reference must be made to the report of the Dean.

Certain special courses connected with those in Finance and Economy it has been thought advisable to abolish. In place of them there has been established (to begin in the present year) a new four-year course, entitled, "The Course in Commerce and Industry." In this course, business and the business life are regarded not as an occupation—as a means of securing a livelihood—but as a profession with duties that reach beyond self; with principles that must be mastered if those duties are to be adequately performed, and with auxiliary sciences or disciplines that cannot be ignored. The large and vital fact that business, as it is called, plays in individual and corporate life, the immense development it has reached in recent years, and the consequent increasing intricacy of its processes, seem ample justification for the establishment of such a course.

A new departure has been made in the method of keeping the College records of the students. So stated, there would seem little, if any, reason for mentioning it here, and I need not describe it. But it has had what may prove to be, and it is to be hoped will be, an exceedingly important result. It has brought about the recognition of a need, long felt, indeed, but never before adequately supplied: a system of Faculty advisers for students. This was found indispensable to this new system of records, that the students might be guided aright in their written statement of the courses they intended to take. But such guidance is hardly conceivable, hardly, indeed, possible, without simultaneous helpful advice in the selection of the constituent subjects of the course. This is particularly true of the courses in Arts and Science, where large freedom of individual selection is open. In the technical courses, the nature of the facts had long ago constituted the heads of those courses the trusted counsellors of their students; but in the courses in Arts and Science there were and are now similar heads unmistakably in evidence. And this was just where such wise and friendly counsellors and helpers were most sadly needed. It is not to be expected that all at once a new habit will develop in our students of going to the men so appointed for counsel as to subjects of profitable study or other matters that are less closely related to the College curriculum. But it is reasonable to hope that the official visit, at the beginning of each term, with the experience thus obtained of questions wisely answered and advice kindly given in cases of doubt, will gradually lead to relations of confidence on the part of the student and interest on the part of the teacher. This relation of confidence and interest once established will furnish a better instrument of control than College discipline—that will be the least of its advantages: it will prove, in the hands of a wise and sympathetic teacher, a most admirable instrument of education, especially in all that teaches character and conduct.

THE GRADUATE SCHOOL.

As the report of the Dean shows, the work of the Graduate School has progressed quietly and uniformly. There are not many matters that call for special notice.

The number of students in attendance continues to show an advance upon previous years. Of the two categories into which they are divided, that of regular students, or students who have been admitted as candidates for a degree, shows slightly the larger increase. We may fairly conclude, I think, that this progress is not due to any accidental or temporary circumstances, but to causes inherent in the character of our work, and likely, therefore, to persist.

Particular attention should be given to the account the Dean has given of the scientific activity of the Senior Fellows. To provide the opportunity and furnish the possibilities for work of this kind was one of the chief purposes had in view in the establishment of these fellowships. It is gratifying to observe that the fruit that was hoped for is forthcoming; and there is no room for doubt that such activity will attract the attention it deserves, will be received as substantial contribution to knowledge, will increase the reputation of the University, and draw to us from year to year a growing body of earnest students.

I may call attention again, as I have in previous reports, to the table which sets forth the various colleges and universities from which our graduate students have come.

The wide distribution of these colleges is pleasingly significant: it speaks for itself, and calls for no further comment.

Of similar interest is the list of graduates, with the titles of their doctor-dissertations, and the positions they have been called to here and elsewhere upon graduation. No better or more efficient missionaries can be desired than a body of worthy graduates so distributed in positions of trust and usefulness in various parts of the country.

A new group has been formed in the Department: that of Archæology and Ethnology. The members of the Group Committee are Prof. Hilprecht, Dr. Clay, Dr. Bates and Mr. Culin. The last-named has been appointed Lecturer in Ethnology and American Archæology. This appointment was made partly to compensate us for the loss sustained in the death of Dr. Brinton, and partly to secure to the Department the valuable services of Mr. Culin, whose growing reputation as a specialist in these studies, as well as his connection with the Museum, must prove of great value. To this group has been transferred the archæological work hitherto scattered through the other groups, to which Prof. Hilprecht and Dr. Bates belonged. It seemed fitting that Archæology, which in the University has been brought prominently before the public by the existence of the Museum, and its valuable collections, should receive corresponding independent prominence in the courses of instruction.

An attempt was made, unfortunately frustrated by untoward occurrences, to institute still another group covering the subjects of Jurisprudence and Political Science. This originated in an offer made by the Department of Law to give certain courses of a scientific character upon subjects within their province under the direction of the Department of Philosophy. With these were to be combined for the purposes of this Department certain other courses in

Economics and Political Science. The plans had been perfected, when the sudden death of Mr. White, of the Law School, who was to have been a member of the group, stopped for the moment all further steps towards their realization. The attempt, however, is noteworthy; if for no other reason, at least as another evidence of the spirit of co-operation that is so wholesomely manifesting itself in the University at large.

In administration, the most important part is the modification that has been introduced in the method of testing and certifying to the qualifications of candidates who are to be recommended for degrees. To this the Dean very properly devotes a large part of his report. Hitherto the candidate was required to pass all his examinations, both in minor and in major subjects, at the same period, and this only after the thesis was completed, handed in and approved. Now it is provided that all the examinations may be passed before the thesis is finished; and the thesis, under provisions in each case to be determined by the Department, may be subsequently completed, if necessary, *in absentia.* This latter provision is intended to make room for the production of such theses as may require preliminary investigation and research in libraries and archives elsewhere, and often widely scattered. Further, the various examinations, instead of being close together, as formerly, may now be distributed: that is, either or both the examinations in the minor subjects may be taken as soon as the requisite amount of work has been covered in each field, the examination in the major being left to the last. This will enable the candidate during the latter part of his course to concentrate his attention upon the new field which alone will characterize him, and proficiency in which alone will give him his standing in the general academic world. The possibility of furthering the completion of his thesis also will give him greater opportunity of concentration upon the one piece of work which is to be his chief credential with the outside world—the one sole evidence to them of the excellence of the training he has had, and of the ability with which he has availed himself of it.

The other modifications are of a formal character, and perhaps of less importance: they have been explained by the Dean in great detail, and he has given the reasons and impelling motives that were back of them. Suffice it to say that the purpose is to secure thoroughness in essence and dignity in form. The written examinations, combined with the onus of responsibility cast upon the examiners, seemed to be the best method of securing the first, which to many is the more important, of these objects. The second, which must, I think, be regarded as also of great value, it is attempted to secure by a presentation of the candidate to the full Faculty by the chairman of the Group Committee in which he has taken his major subject, or by a delegated member of that group. The details as to the method of this presentation are given by the Dean. There are two sides to this. The candidate is thus formally, with due ceremony, introduced into the academic or scholarly world, as officially represented by the Faculty of the University; the Faculty is made acquainted with the approved title (or the title as yet to be approved, would perhaps be more accurate) of the candidate to such introduction. Finally, for the information of the academic world at large, these statements thus given to substantiate the candidate's title to the degree sought, and to give warrant to the Faculty for the recommendation that it be conferred, are to be published *in extenso* and distributed to all sister institutions.

These provisions would seem to secure that amount of

dignified formality which so important an act on the part of the Faculty and an event so serious to the candidate fairly and rightly demand; and further would seem to fix the responsibility that attaches to such a recommendation in a very sensible way first upon the Faculty at large, and finally upon that portion of the Faculty which originates and moves the recommendation.

It is my desire, in this report, to record the thanks the University owes to Dr. N. C. Schaeffer, State Superintendent of Public Instruction, who has so kindly consented to supply temporarily the place of Professor Brumbaugh, absent on public service in Porto Rico.

THE DEPARTMENT OF LAW.

The first-year class has a total registration of 120, being an increase of 29 students as compared with the first-year class of one year ago. The total registration for the entire school shows a decrease of seven, which is due to the fact that the graduating class of 1899 was the last to enter under the old entrance requirements, and was therefore a large class, numbering 114 men. The result is that the school has about regained the number of students which it had before the increase in the standard of requirements of admission. The result of this increase in the standard is shown in the percentage of college graduates who now attend the school. Of the total number of students, 38 per cent are college graduates; of the First-Year Class 41 per cent are college graduates. In addition to these percentages, 20 per cent of the total number and 16 per cent of the First-year Class have attended college for a longer or shorter time, but without receiving degrees; so that more than one-half of the men in the Law School are men who have attended college.

As was to have been expected, this large proportion of college graduates has resulted in an increase in the percentage marks obtained upon the annual examinations. These marks show that of the students from the large universities, the average percentage received was six higher than the average of all the students, while the average of students from all the colleges was five higher than the average of all the students. And the percentage of honors shows also that the college graduates are much more competent to perform their Law School work with success, the first ten men in the graduating class being college graduates. The result justifies the hope that, by increasing the proportion of college graduates by means of increasing the standard of admission to the Law School, the average of work done in the school would be improved proportionately.

It is also interesting to notice that in the Law School there are thirty-five students from States other than Pennsylvania, as against eight from States other than Pennsylvania ten years ago. This shows that the Law School has obtained a high standing throughout the country.

The Faculty has suffered during the last year by the death of Mr. James Parsons, Emeritus Professor of Law, and Mr. Roy Wilson White, Instructor of Law. Mr. Parsons had served as active professor from 1870 until 1897, and was a scholar widely known for his learning and writings. Mr. White was a graduate of the Department in the Class of 1898. He had spent several months at the University of Paris preparing a course upon the Civil and Roman Law, and had just returned from Paris when he was murdered. He was a man of unusual ability, and a very successful teacher, and his loss was deeply felt; and not the least in the proposed Graduate School work.

Perhaps the most noticeable addition to the courses of

the Department was the course in New Jersey Practice, which has been so successful that it has been extended in its scope; and it is now contemplated adding a course in Delaware Practice of the same character.

The Faculty had almost consummated an arrangement by which a course should be given during the year 1900-1901 leading to the degree of Master of Laws, and including Comparative Law, History of the Common Law, the Study of the Problems of Common Law, Government and Public Law. The sudden death of Mr. White, who was to have sustained an important part in the graduate work, rendered the giving of this course impossible during this year, but it is hoped that the course will be given during the following year.

The dedication of the new building of the Department of Law, at Thirty-fourth and Chestnut streets, on the twenty-first and twenty-second of February, 1900, was the most important event of the year; and indeed marked an epoch, not only in the history of the Department, but in the history of legal education in this country. The ceremonies were attended by a large number of eminent persons: Judges, Lawyers, University and College Presidents and Professors, representatives from the Supreme Court of the United States, and representatives from the Universities of Oxford and Cambridge. The dedication ceremonies included a reception on the afternoon of February 21, in the new building, followed by the formal exercises in McKean Hall. Mr. Dickson, chairman of the Trustees' Committee on Law and Legal Relations, presented the building to the Faculty of the Department of Law, the presentation being accepted by the Dean of the Department. Mr. James Barr Ames, Dean of the Harvard Law School, then delivered his address on "The Vocation of the Law Professor." On the same evening, the dedication exercises were continued at a meeting held at the Academy of Music. At this meeting, addresses were made by the Hon. John M. Harlan, on "The Constitution of the United States," and by Sir Charles Arthur Roe, on "The Constitutional Relations of England and Her Dependencies."

On February 22—"University Day"—the guests of the University attended the usual exercises at the Academy of Music. The afternoon of the same day was devoted to receptions in the new Law School building; to an address by Mr. Carson, dedicating Price Hall; and to an address by the Hon. G. B. Finch, of Cambridge, England, on "The Development of the Common Law." The ceremonies closed with a commemorative dinner of 650 persons, given in Horticultural Hall under the auspices of the Law Association of Philadelphia, the Lawyers' Club of Philadelphia, and the Bar Association of Pennsylvania.

In connection with the new building, it is interesting to note that several important paintings have already been placed on the walls, in accordance with the hopes upon which the plan for the building was made. A portrait of Mr. Richard C. McMurtrie now hangs in the McMurtrie Reading-Room. A portrait of Chief Justice McKean, by Vonnoh, is now hanging in McKean Hall. A portrait of Mr. Justice Wilson, first Professor of Law in the University, adorns the main staircase. Judge Wilson's association with this school is further marked by a number of valuable original documents belonging to him, and presented to the University by Professor Carson. These include a letter of President Washington, forwarding to him his commission as Associate Justice of the United States Supreme Court; his original commission itself; his commission as Avocat-General, signed by Louis XVI; and his appointment as Professor of

Law in the University of Pennsylvania. These memorials have been suitably framed, and now hang on the wall of the main hall.

DEPARTMENT OF MEDICINE.

In proportion as Medicine acquires the exactness of a science, greater demands are made upon instructors as well as upon students. Growth in comprehensiveness has made auxiliary to Medicine the sciences of biology, chemistry, physics and mathematics. The activity with which research in medicine is presented in all civilized countries, and the rapid multiplication of publications, have rendered a knowledge of modern languages, especially German and French, more and more a necessity for the Doctor of Medicine. These advances can best be met by a preliminary college training, in which especial attention is devoted to the biological and physical sciences and to modern languages.

The rapid advance in the fundamental branches of Medicine—anatomy, including histology, physiology, medical chemistry and bacteriology—which from their nature and relations to other subjects of the Medical curriculum should rightly be taught early in the course, render it imperative that four years should be devoted to medical studies. The special training required of the teachers of these branches, and the laboratory equipment now recognized as necessary for the proper instruction in human anatomy, physiology, medical chemistry and bacteriology, have tended to render the "literary college" more and more unsuitable for affording adequate opportunities for the prosecution of this work. In order that the medical student may not enter upon his course hampered by an insufficient preparation, it has been decided—beginning with the session 1901-02—to abolish the privilege of admission to advanced class standing on the basis of pre-medical work done in college.

Believing in the correctness of the pedagogic principle which provides for the non-repetition of work already satisfactorily accomplished, the Faculty has adopted a rule exempting students from attending courses upon subjects in which they have given evidence of satisfactory preparation. The provisions of this rule may be expected to apply to certain fundamental subjects—inorganic chemistry, histology and embryology—the natures of which permit of their successful pursuance during a college course. As already indicated, it is the conviction of the Faculty that the more strictly medical subjects, such as medical chemistry, physiology and bacteriology, cannot be dealt with outside of a medical school, in such a way as to render superfluous instruction in those branches as offered in the Medical course.

Four annual courses of instruction, of eight months each, are now generally recognized as representing the minimum period requisite for anything approaching to an adequate preparation in the science and art of Medicine. But so rapid has been the accumulation of exact and minute knowledge concerning health and disease, that it has become a customary practice—and one to be heartily encouraged—for the prospective practitioner of medicine to supplement the knowledge required for the degree of Doctor of Medicine, and qualify himself further by a residence of one or two years in a hospital. Moreover, to this training not a few add some months of post-graduate work in this country or abroad. The superior training which such prolonged study insures is to be strongly commended. But the difficulty is immediately encountered of the very mature age at which young men disposed so to prepare themselves can enter upon the active duties of their profession. This subject has been considered at length in the comments upon Dean Penniman's report. I need only refer to that discussion, and to make note here again of the establishment of a composite year, in which members of the Senior Class of the College may matriculate in the Department of Medicine, and take, in conjunction with certain studies of that year in the College, those belonging to the first year of the Medical course. By this arrangement, a year's time can be profitably saved to the student in the Medical Department whose studies there are preceded by an acceptable preliminary college course.

The signal advances in medical science and instruction, which have been accomplished in recent years, are due in great part to the introduction into medical schools of the laboratory idea. Physiology, pathology and pharmacology, for the proper appreciation of which, only a few years ago, didactic lectures were considered all-sufficient, are now, in large part, taught by means of practical work carried out by the student himself. By this method knowledge becomes a living thing; while the student learns the meaning and use of instruments of precision, and becomes skilled in logical reasoning. But a modern medical school has a second purpose to subserve besides that of instruction alone. It should be a centre of research in the science of Medicine. Nor should research be fostered solely in a spirit of advancement, but also because it insures activity on the part of the instructor, while stirring up increased interest on the part of the student. In order to provide for these pressing necessities in our own Medical School, increased accommodation and facilities are urgently demanded. It is believed that the commodious building, which has been planned for the departments of physiology, pathology and pharmacodynamics, will be begun at an early date. The nature of the work which it is proposed to carry on in these laboratories is of the greatest public interest, as bearing upon the fundamental questions of health and disease, and cannot fail to elicit the support of an enlightened community.

The importance and need of the endowment in this country of medical schools cannot be too strongly emphasized. The wisdom of such beneficence is shown in the results achieved in respect to the knowledge of the causation, prevention and cure of disease, which has been acquired in the last quarter of a century. The brilliant achievements of Pasteur, Koch, Behring, Haffkine and many others, in alleviating suffering, and the many avenues for research which have been opened up by their work in the science of Medicine, should make this field of human activity one of the most attractive to those who are able to contribute towards the beneficent increase of human knowledge. The importance of endowments devoted to the scientific branches of Medicine is exhibited in the achievements of the Pasteur Institute in Paris, which is supported by private subscription and munificence; and hardly less by those of the laboratory in Berlin presided over by Robert Koch, and maintained by the German Government. In this country, where State aid for education is of a limited character, it is incumbent upon the wealthy and generous members of the community to provide for the promotion of American research in Medicine.

Among the means that may be employed to facilitate and stimulate research in Medicine, the creation of medical fellowships should be especially emphasized. The income from such fellowships, by enabling the holders to devote their whole energies to research work, unhampered by the necessity of expending a major part of their time and much of their strength in teaching or other means of securing a livelihood, may confidently be expected to yield fruitful results. No elaborate argument is required to convince those interested in higher education of the fundamental value of this form of endowment. The signal advances in higher education in this country during the last twenty-five years have been in no small measure the result of the creation and maintenance of a large number of fellowships by our universities. Owing to various causes, Medicine hitherto, to a large extent, has been deprived of these advantages; but within the past decade several important fellowships have been founded in different medical schools, and we may venture to hope that the interests of the Medical Department of this University may also be furthered in this way.

The State is capable of fostering medical education mainly in two ways. First, it can afford facilities for the student by contributing towards the maintenance of sound educational institutions and well-equipped hospitals in which opportunities for instruction are offered; and secondly, it can exact a high degree of efficiency before conferring upon a candidate a license to practice medicine. The latter means of insuring to the public the best possible medical service should be jealously guarded by the officers of the State appointed to adjudicate upon the qualifications of applicants. The Faculty is in entire sympathy with every legitimate effort on the part of the Medical Council in its endeavors to raise the standard of medical education in this State.

During the last session the Faculty has tested certain changes in the curriculum of the first two years of medical study, whereby a more perfect sequence in instruction was found to be obtainable. In the present year, as a further modification, chemistry, anatomy, physiology and pathology are being concentrated into shorter periods; while at the same time more hours each week are devoted to their pursuit than in former years. Thus, the practical laboratory courses in pathology and physiology cover nine hours a week for half the term, instead of three hours a week for the entire session. In this way it is believed that two distinct advantages have been gained, inasmuch as the actual laboratory hours have been increased by almost fifty per cent; while at the same time the concentration of thought involved on the part of the student has already produced better results than were obtained under the old system. Moreover, the curriculum is so arranged that inorganic and medical chemistry, histology and bacteriology are given in the first year of the course, in order that pathology and physiology may be followed by the student with more profit in the second year. While it is too early to speak finally upon the value of these changes in the curriculum, the Faculty has good reason for regarding them favorably.

The activity in the colonization of foreign lands by the Caucasian race has emphasized the need of a better knowledge of tropical hygiene. Within the past two years several medical commissions have been sent into tropical countries from England, Germany, America and Japan, to study and report upon the prevalence of disease, and to propose means of rendering these countries more habitable and healthful for Europeans. In addition to this, special schools of medicine for the study of tropical diseases have been established in London and Liverpool. Recent events have so increased the importance of the tropics to the United States that, for the protection of a relatively large proportion of our fellow-citizens, there is urgent need that facilities for instruction in tropical diseases should be supplied by this country. Our Medical School is already equipped with laboratories in which fundamental work in such a course could be given, and for the present it would be necessary only to offer in addition a set of lectures upon the main problems of tropical pathology and hygiene. It is believed that such a course would appeal to medical students in this country, to qualified young men in our new possessions purposing to study medicine, and to practitioners who are planning to settle in tropical climates.

DEPARTMENT OF DENTISTRY.

As was expected by the Faculty of the Department of Dentistry, the increased entrance requirements demanded for admission to the First-year Class of the session 1899-1900 had the effect of reducing the number of matriculates to 158, as compared with the class numbering 183, which was admitted at the beginning of the session 1898-99. Doubtless further increases in the entrance requirements will have a similar effect; but the reduction in the number of entering students occurring whenever increased entrance requirements are demanded, as history shows, is only temporary in character. So soon, then, as the entering class, in the near future, shall have reached the maximum number that can be effectively instructed in the Department, the Faculty should incline towards a gradual increase in entrance requirements, until the requirements for admission shall have reached the standard for admission to the College, the Department of Law and the Department of Medicine, in which identical requirements are at present demanded.

The report of the Dean shows that the Department is continuing to attract students entering to advanced standing from other institutions, and that a marked increase has occurred in the number of students from foreign countries. This latter showing is extremely gratifying, because it is an indication of the esteem in which the instruction given in the Department is held by the dental profession and intending students in those countries.

The Dean of the Department, in referring to the differences in the character of high school instruction as, to some extent, defeating one of the purposes of the increased entrance requirements, viz., "that of selecting students whose education best adapted them for professional instruction," is quite right in his conclusions; but possibly this difficulty might be overcome by differentiating between high schools.

Within recent years the studies which should be pursued by students of modern Dentistry have so increased, not only in number but also in extent in dental schools of the highest class, that a rearrangement of the curriculum appears to be necessary. Whether a rearrangement of the curriculum of the first year of the course, as indicated by the Dean as being in progress at present, will suffice; or whether a rearrangement of the studies of the subsequent two years will also be considered a necessary incident to the requirement of a great number of specified studies in the course; will no doubt be seriously considered by the Faculty. It is quite possible that the increase in the number and extent of necessary studies demanded by the requirements of modern Dentistry may compel, in the near future, the addition of a fourth year to the course.

It is gratifying to observe the decided improvement, referred to by the Dean, in the character of the technical work which is being attained through the extension and elaboration of the laboratory method of instruction. As so much of the teaching of Dentistry is technical in character, a large staff of laboratory instructors is necessary adequately to carry out this work, and this will more and more require the services of instructors who can devote their entire time to it.

DEPARTMENT OF VETERINARY MEDICINE.

Since the presentation of the last report of the Dean of the Department of Veterinary Medicine, the financial support which might be expected, especially from that part of the public interested in the raising of the lower animals for profit, and also from those members of the community who are deeply interested in animals from a humane point of view, has not been received. I say this with much regret. No professional department of the University is more acutely in need of endowment or direct contributions for support than the Department of Veterinary Medicine. Were it not for the generous contributions which are annually made by the children of the late J. B. Lippincott, one of the founders of the Department, the school would indeed be in sore straits for maintenance.

The high educational standard of the Department is continued, and attracts the better class of students of veterinary science. It is pleasing to know that the educational qualifications exhibited by the members of the entering classes are annually and progressively increasing. As stated by the Dean, more than one half of the present classes are graduates of colleges or high schools.

The educational training which the students of this Department receive peculiarly qualifies them for teaching positions in institutions where veterinary science is taught; and for positions in experimental stations where researches in agriculture and in animal husbandry are conducted, as well as for positions in the Bureau of Animal Industry of the National Government which are concerned in the inspection of meat. The Dean in his report refers to appointments of the character described as having been accorded to several recent graduates of this Department.

While the work of the veterinarian, to a great extent, may be directed towards the curing of disease among the lower animals, it is of equal if not greater importance that his knowledge and skill be employed towards the prevention of disease; and especially the prevention of the plagues too common among the lower animals, and the diseases that are transmissible from the lower animals to man.

It is to be hoped that the earnest plea of the Dean of the Department, that provision may be made so that the undivided time and energies of the members of the Faculty may be employed in the Department, will be fulfilled. This object can be accomplished only by the generous liberality, in the way of endowments and contributions, of friends of the Department.

THE UNIVERSITY LIBRARY.

The report of the Librarian is of interest as showing the present greatly increased value and importance of the Library of the University. This growth in numbers of volumes, and still more of readers, is undoubtedly largely due to the completion of an entirely new revised dictionary card catalogue, both of subjects and authors, prepared at the expense of one of our constant and liberal benefactors. In this way, at last, every student can find every book on any subject. Then, too, the rearrangement of the Library with alcoves, each bearing the name of a benefactor, and containing reference and other works on some subject specially connected with his work, and with a separate reading desk for each student in an inner room, and a goodly array of periodicals (over a thousand now being received and put on file in the Library), and facilities for readers, and the ready service of books in the outer library, has attracted a largely increased number of students. Every effort is made by the Library staff to meet promptly all demands, and students and professors alike have testified to their satisfaction with the increased facilities provided for their use. Liberal gifts of money for special collections and of books have facilitated the work. Several of the recent College classes, following the example of earlier classes, have made gifts of money or books for the Library as their memorial; and it is proposed to put up in the Library a tablet, containing a list of all classes, and of individual students and graduates and others who have made the Library the recipient of their benefaction, as an inspiration to others to do likewise. Then again: the opening of the upper rooms of the Library building [vacated by the removal of the Museum to its splendid new building] to the seminaries for post-graduate and other students, has provided space in these rooms for special libraries; and instructors and students have testified by their greatly increased use of these books their acknowledgment of this important addition to the facilities for continued work. Many of these rooms are kept open to a late hour every day, and not an evening passes without the attendance of earnest students. The same thing is found in the greatly increased use of the valuable Library of the Law Department; and this of itself is an addition that gives to the students and instructors a renewed impulse to additional study, and to the preparation of work that requires the largest use of the Library. Along with this growth of the Library of the University, and of its proportionately increased usefulness comes, however, the renewed demand for an endowment that shall provide for its annual requirements. First, for its good administration, with capable experts in charge of each of its many branches; and second, for the regular addition of every book needed by the students and teachers in all the broad fields of science and instruction now included in the work of the University.

Books are largely the tools required; and the Library of the University should be able to put on its shelves promptly after publication every new book of value, and to complete the literature of every branch of learning. To do this even imperfectly to-day is a heavy tax on the limited resources of the University, and on its few steadfast benefactors.

The example set by the College classes, and by the Dental Department, may well commend itself to successive graduating classes. There could be no better memorial than a long and steadily growing list of benefactions of this kind to the Library, for it is used every day by a largely increasing number of students and teachers of the University —in ever-growing proportion, and with steadily advancing facilities. Then, too, the Library of the University may well perpetuate the names of those who have done honor to the University, just as it now points with pride to the collections that bear the names of Henry Reed, and George Allen, and Charles Krauth, and Charles Stillé, and Fairman Rogers. A fixed income, either from endowment or from annual contributions, or from both, is one of the first conditions for the growth of the Library of the University in due proportion to the growth of the University, and all of its departments. The main Library, with its reading rooms and alcoves, the seminary libraries, the Medical, Law, Chemical and other technical libraries, constitute one great whole that needs administration under one head, and with the largest facility for the use of both students and instructors in their own work-rooms, as well as for the freest circulation: but there must be an endowment and an annual income large enough to provide for the regular increase of the Library, and for its staff of experts.

The report of the Librarian shows that while much has been done from individual gifts of money and books, more is yet needed to maintain the steady growth of this storehouse of learning, from which all the branches and all the members of the University are drawing so freely of the treasures of the past, and of the growing wealth of books that represent the present conditions of knowledge to-day. One of the most useful functions of the Library of the University is as a medium of a large system of exchanges with other libraries, and with the growing productiveness of other universities and all learned bodies. To maintain and increase this the Publication Fund of the University needs to be strengthened and maintained, so that the share of our own University in literary activity may be represented in its own Library, and in the libraries of other universities. The steady addition of the books, pamphlets, theses, periodicals, etc., that are daily issued keeps us in touch with the world's growth in learning and knowledge; and the Library of our own University must be one of our foremost cares.

Those who provided the building now devoted exclusively to the Library of the University builded better than they knew. Already it is full to overflowing with the wealth of past ages, and with the products of the intellectual activity of to-day. Departmental libraries furnish to our teachers and students the books needed, and liberal gifts and benefactions continue to enrich our stores. Not only must we continue to collect books, and to increase our long list of scientific and literary periodicals, but we must make them accessible to every student and provide every facility for their free use.

The large gifts to the Library, made by publishers and authors and collectors, show how much its usefulness is appreciated; but properly to show our appreciation of this generosity, the University must be put in a position to make due acknowledgment by having every book in place, and so catalogued that it will be easily accessible to every reader.

To do this requires a large force of expert employes, and the outlay for this purpose must necessarily be large. Every library, to do its work to the best advantage, needs an income sufficient for its proper and liberal administration, and for its due increase by the purchase of books, old and new. Now that the Library has a complete catalogue, it starts out well equipped for its work; but that work needs steadfast support to enable the Library to keep pace with the growth of the University, and to meet the requirements of its teachers and students.

What the libraries of other universities are, in the Old World and in the New, may be largely ascribed to the splendid benefactions that have been made to them. The Library of the University of Pennsylvania has benefited by many individual gifts of value; and the long list of those who have thus perpetuated their names, and those of men and women who have done the University great service, is something of which we may all be proud. We still wait, however, for some one who, by a liberal endowment, shall put our University Library on an independent foundation, thus ensuring its future, and relieving the straightened resources of the University from the growing burden of its proper maintenance and increase. What our Library is to-day is due largely to individual gifts; and what it may become, and what it ought to be to hold its proper place among university libraries, might well commend it to the generosity of those who desire to be enrolled among our great benefactors.

THE FLOWER OBSERVATORY.

Possessing its present staff and equipment, the University was able, during the last year, to take that part in an important scientific observation which, a few years ago, would have been impossible. Although our own Rittenhouse and Smith laid so early the foundations of astronomical science in this country; and although, in later years, our honored Dr. Kendall was among the most distinguished American astronomers; the absence of any adequate equipment shut the University out of that field in which others were making substantial contributions to astronomical science. This year a properly equipped party, led by Professor Doolittle, was sent to Barnesville, Ga., and was largely successful in making the observation of the solar eclipse which had been thoughtfully predetermined and provided for. But this was one of the special and rare occurrences. An examination of the Director's report will reveal the vast amount of painstaking work, conscientiously and faithfully pursued, night after night, and year after year, not with the expectation of any startling or brilliant discovery, but only to add fact by fact that which may increase our knowledge of the heavens; or, more frequently, to make more exact the knowledge we already have. All over the world this patient work is going on, and we may now take pride in the thought that our own Observatory is doing its share in these lofty researches. It is also to be noted that instruction in Astronomy is held to be the first purpose of the Observatory, and that the more fascinating work of research is not permitted to thrust aside that discipline by which the astronomers of the future are to be trained.

THE BOTANIC GARDEN AND VIVARIUM.

Apart from its educational value, the Botanic Garden would be worth maintaining on account of its attractiveness as a feature of the University Grounds. The space appropriated to it is necessarily limited, but the judicious landscape gardening needed for the streets vacated by the city largely increases the area available for its purposes. The new greenhouses and the grounds have been enriched by generous donations of plants, so that there is now a large store of living specimens, sufficient not only for the work of the classes in Botany, and for the Botanical Society of Pennsylvania, but also for gifts of operative material to the Girls' High and Normal Schools and other institutions.

The Vivarium has proved to be well adapted to the purpose for which it was built; and it is gratifying to learn from Professor Conklin's report that it has been eminently successful in dealing with an especially difficult problem. The maintenance of a marine aquarium at such a distance from its source of supply involves many difficulties; and our warm thanks are due to Messrs. Peter Wright and Sons for aid in overcoming the greatest of these, and enabling us to start the aquarium with natural sea-water. This once secured, and its proper aeration ingeniously provided for, there remains only the question of temperature; and we find that the building is of such location and structure as to favor the aquarium in that respect. It will evidently be necessary to supply and manage the fresh-water aquaria in a similar way. Many of the principles of Zoölogy are best learned by the study of the aquatic forms of life; and with these aquaria, successfully maintained, the students are well provided all the year round with the material for such study.

THE WISTAR INSTITUTE.

This is a department which makes no draught upon the University's resources, and adds no weight to the responsibilities of administration. With its ample endowment, its efficient Board of Managers, and, above all, the vigilant and constant care of its founder, it is doing a great deal more than taking care of the old Wistar and Horner Museums. This it is doing, and enlarging them by constant accessions of the greatest value. But apart from this, it is steadily developing a scientific idea of the functions and methods of a modern museum. The forms of cases and trays in which specimens are kept have been carefully planned by the Director and his Assistant, and are not yet used elsewhere. The system under which the specimens are displayed has been carefully thought out, with a view to making them instructive in the minutest degree. It would be impossible to carry out such plans without large buildings and large resources. These General Wistar has provided, and is still increasing; and while it will take a long time to complete and display the anatomical series that is planned, enough has already been done to indicate the full scheme, and to be even now of distinct use to students of Anatomy.

DEPARTMENT OF ARCHÆOLOGY AND PALEONTOLOGY.

It was your pleasure during the past year to convey to Messrs. Clarence H. and E. W. Clark your congratulations upon the success of the Archæological work of the University, which from its beginning has received their earnest thought and liberal support. Similar congratulations may be extended to all who have shared in it, whether explorers in the field, curators of collections, donors of collections, or contributors to its funds. The section of the building now in use has not only been found admirably adapted to the purpose of its construction, but has attracted the admiration of scientific men here and abroad, and will doubtless be a model in one feature or another for other important museums. The results of last year's field work by Messrs. Furness, Hiller and Harrison in the Naga Hills of Assam, and of Mr. Culin among the Indian tribes of the Northwest, have enriched the ethnological collections; while that of Prof. Hilprecht in Babylonia has exceeded those of all previous years in the extent and value of its finds, and in their archæological importance. It will take many years of decipherment and collation to gather all the fruits of these Babylonian records; but Professor Hilprecht is so expert in handling the subject of his life-work that, while the excavations and collections are going on, he is able to determine point after point in the history, the daily life, and the religious thought of the dwellers by the Euphrates through many centuries. Apart from these results of our own explorers are those acquired through the co-operation of our Archæological Department with the Egypt Exploration Fund, both in Egypt and in Abydos, and with the American Exploration Society. Through these we have acquired not only sculptures, bronzes and sarcophagi of great importance, but such a share of the recently-found Egyptian Papyri as will make up, to some extent, the deficiency of our Museum in these manuscripts. Individual collectors have added to their deposits or collections already presented; and the Museum is so steadily increasing in value and interest that we may look forward to the time when the buildings must be extended in accordance with the general plan which has been adopted. Meanwhile, it is gratifying to know that the earnest workers in this field are adding to their membership, which, during the past year, has more than doubled. The larger this membership, the more efficient the work of the Department; and when we consider its intrinsic interest, its great value in public education, and the honor which it reflects upon the University, we cannot too strongly advocate its liberal and continued support. It should be remembered, too, that as a Museum it is freely open to the entire community, and will in time come to be recognized as one of the chief ornaments of the city.

UNIVERSITY HOSPITAL.

The University Hospital continues its beneficent work with an efficiency that increases every year. The improvements in equipment which are noted in the report of the Secretary have been chiefly those provided by liberal friends, notably "The James Dundas Operating Room for Private Patients," and make practicable the treatment of patients by the best methods and under the safest conditions. But the enlarged facilities of the Hospital involve certain increases in expenditure. The fact, too, that under the reorganization of the courses in Medicine and Surgery in the Medical Department a greater amount of practical work will be required, and that for this there must be more clinical material, demands a still further increase in its resources. When I speak of clinical material I am not thinking of the patients in the Hospital as only, or even primarily, there for that purpose. For I am assured by the testimony of our most enlightened physicians that clinical instruction is accompanied with the very best care of the patients; where the physician or surgeon prescribes under the eager scrutiny of the alert medical student, and when he must be even surer than usual that the nurses and attendants are accurately carrying out his directions. It is a direct advantage to the public, as well as to the School, when members of the Hospital staff have the privilege of giving bedside instruction. But just now the Managers of the Hospital are confronted by another difficulty. The temporary financial assistance which they have for a few years received from the Medical Department will in all likelihood be cut off by its own needs, under conditions brought about by the advance in the character of its work. A distinct effort must therefore be made by the Managers to increase the endowments of free beds, and the gifts for current expenses, if the Hospital is to maintain its efficiency. The University Hospital is well managed, and deserves the hearty co-operation of all who feel compassion for the sick and suffering to whom the Hospital ministers. I note with pleasure a new system which provides a graduate nurse to supervise the work in each ward, as safeguarding that service on which sometimes as much depends as on the skill of the physician; and also that the home comfort of the nurses and nurse students has been more fully cared for.

SPECIAL EVENTS OF THE YEAR.

Within the year under review there have been many important occurrences. The purchase of what has been known as "the Foulke and Long property" added important buildings and a large area (110,000 square feet) to the estate of the University. The purchase price of this property was $112,500. At another place in this report I refer more at length to this acquisition, and to the purposes for which it is destined.

On February 21 and 22, 1900, the new building of the Department of Law was dedicated, with impressive ceremonies. The Dean of the Law Department refers to this epoch in the history of the school. The twenty-second of February was observed as a high "University Day," with more than usual feeling and attention. His Excellency the Chinese Minister delivered the oration. His subject was "The Proper Relations of the United States with the Orient." The full proceedings of this University Day will be published in connection with the dedication of the new building of the Department of Law. I think that we can all agree that no other function in the history of the University has brought together so many distinguished persons from so many different localities. The memories of the day, the guests of the University, the honorary degrees conferred, the public interest, the presence of representatives from Oxford and Cambridge—all made the occasion one which will long be remembered in our annals.

Within the year there have also taken place two other events of prominent and great importance. On December 28, 1899, the new Museum of Science and Art was formally opened. The Provost of the University delivered the address. Upon the same occasion a gift of $50,000 was presented, as the endowment by Mrs. William Pepper of "William Pepper Hall,"—a memorial to my predecessor.

On February 13, 1900, the corner-stone of the Memorial Tower and Gateway, in memory of the sons of Pennsylvania who fought in the Spanish-American War, was laid by the General of the Army. This Memorial Tower and Gateway is the great entrance to the dormitory system, and, as you know, is being erected in conjunction with an extension of the present dormitory system. Indeed, as I write this report, both Memorial Tower and new Dormitories have been completed. The cost of this new construction, including furnishing, is in the neighborhood of $240,000. There has been, therefore, expended upon that part of the Dormitory plans so far completed, over $600,000, and we have now room for 525 students. It is not to be forgotten that we are building upon a general plan, completed before the first section of the Dormitories was begun in 1895. The plan includes a University Chapel, a University Dining Hall and additional Dormitory houses; affording, when the last shall have been built, room for about one thousand students. I need not say that it is our earnest wish to be enabled to go on with as little delay as possible with this work—the Chapel, the Dining Hall, the other Dormitory houses. I am sure that great support in the future will come from those who have been in residence in these recent years, as well as from those who will hereafter be in residence in the Dormitories.

So much attention has been called to, and enthusiasm aroused by, the last Nippur expedition of the University that I need not refer here at any length to the facts of the case. This last expedition, under the direction of Professor Hilprecht, has found at Nippur the library of the Great Temple. Perhaps no work of the University has brought so much recent renown to the City of Philadelphia and the State of Pennsylvania, at home and abroad, as the historical results connected with the patient and prolonged excavations of this ancient city.

Not the least in the events of the year is to be noted the rapid formation of Alumni Associations. It is quite true that alumni of the University of Pennsylvania are residents of almost every town and city in the United States; but hitherto they have not been grouped together in alumni organizations. Our rapid growth in the last few years has aroused feelings of pride and affection in our graduates, and in many parts of the United States new Associations have been formed. I have been present at a number of "Reunions"; and it will be a great pleasure to me if, in the course of a few months, a convenient itinerary could be arranged by these Societies, so that I might visit all of them with the least loss of time.

The matter of a complete organization of our alumni has been discussed at length by the Vice-Provost and the Deans of the several Departments, in association with myself; and we hope to propose a plan by which, without detriment to any departmental society, we may yet exist as one University body, belonging to one Alumni organization, with means of intercommunication which have never before existed. This seems to all of us who are responsible to you for the administration of the University one of the most important of our internal affairs.

CONCLUDING STATEMENTS.

The cash receipts between September 1, 1899, and August 31, 1900, from donations from all sources for University purposes, are reported by the Treasurer to have been $531,154.39. And as the year under review is the closing academic year of the century, it may be well to state that the aggregate of contributions paid in cash, during the present administration, namely, from June 9, 1894, until August 31, 1900, is $3,335,473.99. In addition thereto the Treasurer has in his safekeeping special subscriptions, payable in annual instalments in the coming few years, and binding upon heirs, executors and assigns, to the amount of about $675,000.

There may be an historical value in making now a record of these facts, so that our successors in administration one hundred years hence may have a basis of comparison.

I will also record at this time the fact that at the last Commencement, held in the Academy of Music, June 14, 1900, degrees in course were conferred upon 551 persons: 104 in the College, 82 in the Department of Law, 26 in the Graduate School, 180 in the Medical School, 146 in the Dental School, 11 in the Department of Veterinary Medicine and 2 honorary degrees. One hundred years ago the degree of Bachelor of Arts was conferred upon five members of the graduating class, and the degree of Doctor of

Medicine upon ten of the class then graduated in Medicine. I place side by side the names of the Trustees of the University of Pennsylvania in 1800, and the Trustees of the University of Pennsylvania upon August 31, 1900:

1800.

Hon. Thomas McKean, LL. D.,
Hon. Edward Shippen,
Rt. Rev. William White, D. D.,
George Clymer,
Thomas Fitzsimons,
Rev. Robert Blackwell, D. D.,
Hon. William Bingham,
Robert Hare,
Edward Burd,
William Lewis,
David Hayfield Conyngham,
Jonathan Bayard Smith,
Frederick Kuhl,
James Irvine,
Charles Pettit,
Jared Ingersoll,
David Jackson, M. D.,
John Bleakley,
Edward Tilghman,
Joseph Borden McKean,
Alexander James Dallas,
Joseph Ball,
William Rawle, LL. D.,
Benjamin Rawle Morgan.

1900.

Hon. William A. Stone, Governor of the State of Pennsylvania, President, *ex-officio*,
Frederick Fraley, LL. D.,
William Sellers,
John Vaughan Merrick,
Richard Wood,
Silas Weir Mitchell, M. D., LL. D.,
Charles Custis Harrison, LL. D.,
Rev. George Dana Boardman, D.D., LL. D.,
Horace Howard Furness, Ph. D., LL. D., Litt. D. (Cantab),
Wharton Barker,
Samuel Dickson,
John Clarke Sims,
Hon. Samuel Whitaker Pennypacker, LL. D.,
Rt. Rev. Ozi William Whitaker, D. D., LL. D.,
John Barnard Gest,
Joseph Smith Harris,
Walter George Smith,
William West Frazier,
Morris James Lewis, M. D.,
Joseph George Rosengarten,
Randal Morgan,
James McCrea,
Samuel Frederic Houston.

You will observe that, upon August 31, 1900, there were two vacancies existing—one arising from the resignation of Mr. Charlemagne Tower, and the other from the death of Dr. J. M. Da Costa.

As the work of the new century opens upon us, I will state several University needs, which, counting upon our recent experience of the widening public interest, we may confidently expect will soon be realized. In my last report I referred to the necessity of the erection of laboratories for the distinctly medical subjects of Pathology, Physiology and Pharmacology. It gives me pleasure to be able to say that distinct progress has been made within the year towards the accomplishment of this important undertaking. I am sure that nothing has been lost by the delay of a few months. A proper site has been selected, abundantly large for the future growth of the Medical Department. The plans have been carefully studied by a committee of the Faculty; and Messrs. Cope & Stewardson, the architects, are perfecting their drawings. No one of us can count upon the continuance of life or health; but if these blessings are vouchsafed to us, I shall be very much disappointed if work upon the new laboratories of Medicine shall not have been begun in the spring of 1901.

It is true that but a small portion of the necessary money has yet been subscribed, and it is also true that the total expenditure will not be less than $500,000; but we shall be able, I doubt not, to raise this fund in the State of Pennsylvania.

I have already referred to the purchase of the property between Thirty-third and Thirty-fourth streets, bounded on the south by Locust street (since vacated by the city); and I have already announced to you a subscription of $250,000 for a Physical Laboratory, to be established upon the western half of this new property. The University is now, therefore, well provided in Chemistry and Physics.

I am sure that we all recognize the duty of materially strengthening, at the earliest possible moment, the Engineering departments. Other professional schools have been placed on an excellent footing, and their relative rank is distinctly higher than that of the Engineering departments. The favorable elements in the State of Pennsylvania for engineering schools of the first rank are evident. Its two great cities are rapidly growing in population, as is the State itself. The manufacturing interests represented in these cities and surrounding terrritory are vast. These require an unusually large proportion of technically trained assistants, both in their operation and in designing their products—for example, such industries as the steel industry, bridge-building, ship-building, locomotive-building concerns, machine shops, etc. I need not refer to the great transportation interests of the State.

Our needs are: (1) A strictly first-class "Engineering Building"; (2) ample laboratory and shop equipment, avoiding all needless duplication in different departments; and (3) of course, a substantial increase in the teaching staff. It is estimated that the cost of this building and equipment will not be less than $500,000.

In anticipation of your early action, Professor Spangler and Professor Marburg have prepared the necessary plans; and the opportunity of service which the University will be able to render, when so equipped, will in due time be submitted to the many who are naturally interested in all that relates to Mechanical, Electrical, Civil and Mining Engineering.

The eastern end of the lot, to which reference has been made above, and which, in its entirety, is two and one-half acres in extent, has been reserved, for the time being, for a University Gymnasium. A committee of Alumni is making an effort to raise the money. It is almost grotesque that a University of the rank of Pennsylvania should be, practically, without opportunities for physical education—that is, that education of the body without which so many of our best young men fail in achieving success.

It is perhaps better not to distract public attention from any single University need by speaking of too many; but I must refer to one or two more. I feel so keenly the urgent need of money to carry on our publications that I venture to add a word in this direction, hoping it may reach the ear of some one able to forward this side of our University activity. It is as essential that scientific work once done should be published, as that the products of manufacture should be put upon the market. Of what use or avail all of our experience, equipment in the way of apparatus, libraries, etc., if the scientific results cannot be placed before the scientific world? The professors and their associates can do the work, but they are not able, after spending time and often much money upon the investigation, to then pay for publication. One of the most serious wants, therefore, is an endowment for our publications. We now own all the stock of the "University Press," so that we can publish under our own name. It is not necessary for me to call to mind that our reputation rests not only upon our material growth, but upon our place in the scientific effort of the world. If we could count upon an income of $4,000 a year, we would be able to do a good deal towards publishing our own original work, and towards permanently maintaining a "Bulletin" of the University. I make this appeal out of a heart full of concern, hoping that some one may be found who will come to our relief in this regard.

Finally, in so far as our important needs are concerned, I wish to call particular attention to a fund from which retiring allowances may be voted to members of the teaching body. This fund was started by our late colleague, Dr. Da Costa, a few weeks before his death, and his will contained a bequest of $5,000 towards the principal sum of such a fund. The University ought to be in a position to vote a retiring allowance to any one of its teaching body who has served for a definite number of years, and who has attained an age limit to be hereafter considered and determined.

In closing this report, it is not easy to find any cause except for thankfulness. The only discouragement under which the University has suffered resulted from the failure of the State of Pennsylvania to make any appropriation at the last meeting of the Legislature, except to the Hospital. You have therefore been obliged, in the past two years, to sustain through private gifts the entire burden of maintaining and advancing a great and indispensable work. "The University has lost, to a great extent, the support and strength it ought to have derived from that most vigorous of Commonwealths whose name it bears." I venture to express the assurance that the State hereafter will be willing —as she may be able—to provide in some reasonable measure for the needs of the University, in accordance with her Constitutional requirements. At the same time, I sincerely hope that we may never call upon the Legislature except to supplement that private benevolence upon which the future of the University must in the largest measure depend.

I cannot end this Annual Report—the last report of the Nineteenth Century—without feelings of solemnity. It is concluded with a heart full of gratitude, and with a mind full of confidence—in which emotions I am sure that all of you deeply share.

I am, very respectfully,

Your obedient servant,

CHARLES C. HARRISON,

Provost.

APPENDICES.

APPENDIX I.

PERSONAL CHANGES.

BY DEATH.

James Parsons, A. M., Emeritus Professor of Law.
Roy W. White, LL. B., Fellow in Law.
John Ashhurst, Jr., M. D., LL. D., Emeritus Professor of Surgery.
Jacob M. Da Costa, M. D., LL. D., Trustee.
Alfred Stille, M. D., LL. D., Emeritus Professor of the Theory and Practice of Medicine and of Clinical Medicine.
Samuel P. Molenaer, Ph. D., Instructor in French.

BY RESIGNATION OR EXPIRATION OF TERM.

John Ashhurst, M. D., LL. D., John Rhea Barton Professor of Surgery, and Professor of Clinical Surgery.
Francis R. Packard, M. D., Instructor in Laryngology.
James H. McKee, M. D., Assistant Demonstrator of Physiology.
Robert P. McReynolds, Assistant Demonstrator of Anatomy.
Robert S. J. Mitcheson, M. D., Assistant Demonstrator of Anatomy.
Joseph P. Tunis, M. D., Assistant Demonstrator of Anatomy.
John P. Arnold, M. D., Demonstrator of Physiology.
Henry R. Wharton, M. D., Demonstrator of Surgery.
Richard H. Harte, M. D., Demonstrator of Osteology.

APPOINTMENTS.

J. William White, M. D., Professor of Clinical Surgery.
Thomas R. Neilson, M. D., Assistant Demonstrator of Surgery.
J. Bird Moyer, Ph. D., Instructor in Chemistry.
Merrick Whitcomb, Ph. D., Instructor in European History.
Lewis F. Pilcher, Ph. B., Lecturer in the History of Architecture.
Georg M. Richter, Assistant Instructor in German.
Cassius H. Watson, Class Demonstrator and Laboratory Assistant in Biology.
Willett L. Hardin, B. S., Ph. D., Instructor in Technical Chemistry.
Frederick M. Mann, C. E., M. S., Instructor in Architecture.
H. Donald Tiemann, M. E., Instructor in Mechanical Engineering.
Arthur R. Bauder, B. S., A. M., Instructor in Physics.
A. Ferree Witmer, M. D., Lecturer in Pathology.
Harry S. Pratt, Ph. D., Lecturer on Transportation and Commerce.

GENERAL.

George Dana Boardman, D. D., LL. D., Lecturer on Christian Ethics.
Edgar F. Smith, Ph. D., Sc. D., Vice-Provost.

IN THE COLLEGE.

George F. Barker, M. D., Sc. D., LL. D., Emeritus Professor of Physics.
Edward S. Meade, Ph. D., Instructor in Industry and Commerce.
Frederick A. Cleveland, Ph. D., Instructor in Finance.
William McClellan, B. S., Instructor in Physics.
Stewart M. Marshall, B. S. in E. E., Instructor in Mechanical Engineering.
John C. Hockenberry, Lecturer in Pedagogy.
Rev. Florian J. C. Vurpillot, B. D., Instructor in French.
James Gillinder, Jr., B. S. in Chem., Instructor in Chemistry.
Orville P. Phillips, Ph. M., Assistant in Biology.
William E. Lingelbach, A. B., Instructor in European History.
Thomas Nolan, M. S., Ph. B., Instructor in Architecture.
William H. Allen, Ph. D., Instructor in Public Law.
Hon. Nathan C. Schaeffer, Ph. D., Lecturer on the Science and Art of Education.
Thomas R. White, B. L., LL. B., Lecturer on Business Law.
Charles F. Osborne, Instructor in the History of Architecture.
Ellwood C. Parry, M. L., Assistant Instructor in German.

IN THE LAW DEPARTMENT.

Edgar S. Shumway, A. M., Ph. D., Lecturer on Roman Law.

IN THE MEDICAL DEPARTMENT.

Richard M. Pearce, M. D., Demonstrator of Pathology.
William F. Hendrickson, M. D., Assistant Demonstrator of Pathology.
Colin C. Stewart, Ph. D., Demonstrator of Physiology.
J. Francis Walsh, M. D., Assistant Prosector to the Professor of Anatomy.
Brooke M. Anspach, M. D., Instructor in Clinical Gynæcology.
George N. Richmond, M. D., Assistant Demonstrator in Anatomy.
Thomas H. Evans, M. D., Assistant Demonstrator in Anatomy.
Edward Lodholz, M. D., Assistant Demonstrator of Physiology.
Walter S. Hanley, M. D., Assistant Demonstrator of Physiology.
Arthur F. Coca, A. M., M. D., Assistant Demonstrator of Pathology.
Richard F. Woods, M. D., Instructor in Clinical Gynæcology.
Robert N. Willson, Jr., A. B., M. D., Instructor in Physical Diagnosis.
John M. Cruice, M. D., Assistant Instructor in Clinical Medicine.
Henry K. Pancoast, M. D., Assistant Instructor in Clinical Surgery.
Joseph Walsh, M. D., Instructor in Clinical Medicine.
Richard C. Norris, M. D., Demonstrator of Obstetrics.

J. William White, M. D., John Rhea Barton Professor of Surgery.
Edward Martin, M. D., Professor of Clinical Surgery.
Charles H. Frazier, M. D., Professor of Clinical Surgery.
Frederick H. Howard, M. D., Assistant Demonstrator of Pathology.

IN THE DEPARTMENT OF DENTISTRY.

George G. Milliken, M. D., D. D. S., Assistant Professor of Operative Technics.
Matthew F. Cryer, D. D. S., Professor of Oral Surgery.
A. DeWitt Gritman, D. D. S., Demonstrator of Prosthetic Dentistry.
Lewis W. Darlington, D. D. S., Demonstrator of Operative Dentistry.
William L. Ellerbeck, D. D. S., Demonstrator of Operative Dentistry.
John A. McClain, D. D. S., Demonstrator of Operative Dentistry.
Joseph G. Hickey, D. D. S., Demonstrator of Operative Dentistry.

DEPARTMENT OF PHYSICAL EDUCATION.

Esther Kuhn, Instructor in Gymnasium for Women.

THE LIBRARY.

Susan W. Randall, Assistant Librarian.

APPENDIX II.

REPORT OF THE DEAN OF THE COLLEGE FACULTY.

To the Provost of the University;

SIR:—I have the honor to present to you a report of the College during the year 1899-1900:

CHANGES IN THE FACULTY.

At the close of the year Professor Barker resigned the chair, which he had held for so long a period, and the following extract from the minutes of the Faculty of the College expresses the feeling with which the news of his resignation was received:

"Professor Barker's term of service covered a period "of twenty-eight years—years which were among the most "critical and important in the history of the University. He "participated in the movements which have given the Col- "lege its present enlarged scope, and have brought it, in char- "acter and aims, in accord with the present conceptions and "requirements of scientific training. As an instructor he "was clear, forcible and impressive, in putting before his "classes the facts and controlling ideas of his subject.

"As a student and man of science, Professor Barker has "moved our admiration by the quickness and penetration "with which he seizes upon the essential features in the most "recent developments and discoveries: by his instinctive in- "sight into their meaning, and appreciation of their bear- "ing upon our conceptions of the nature and working of "physical phenomena: by his power of unfolding in lucid "expression the difficult achievements of his fellow-laborers, "and of replacing their strictly technical reasoning by simple "but striking and adequate considerations and experiments, "so that even the layman can follow them intelligently: by "the vigilance that allowed no published work to escape his "notice, or, once seen, to slip from his memory: by his firm "grasp of the historic development of his science, and of the "illuminating skill with which he set it forth in rapid retro-"spect. It has been given to him, too, to be able to add to "his exposition, whether of scientific principle or scientific "story, that personal touch which can only come from close "acquaintance and intimate friendship with the leading "names of physical science the world over.

"In the councils of the College we have found Pro-"fessor Barker an advocate of all true advancement, a jeal-"ous and vigorous guardian of the rights of the subject of "Physics as an important factor in college education.

"In personal intercourse he is a charming companion; "and such as have heard him informally discoursing in a "social group upon his favorite subject will not soon forget "either what they heard, or the speaker himself: so easy, "so clear, so entertaining was he.

"He has ceased to be our colleague, except in memory; "the College and each member of its Faculty still claims the "honor of calling him a friend."

Continued ill-health having rendered it necessary that leave of absence be granted Professor Barker during a large part of the year, the direction of the courses in Physics devolved upon Professor Goodspeed, whose efficiency and success in the performance of the numerous duties of the position are worthy of special commendation.

The illness of Professor Laird, which extended over several months, rendered necessary certain rearrangements of the work in the course in Architecture. The executive work of the department was placed, during Professor Laird's absence, in the charge of Mr. Pilcher, whose energy and success, both as a teacher and as an administrative officer, has gained for him recognition not only in our own College, but also in other institutions. It was with great regret that his resignation was accepted at the end of the year; but the regret was mingled with pleasure, inasmuch as he left us to take charge of the work in art at Vassar College, with the title of Professor of Art. Mr. Pilcher's work with us will, however, continue, as he was elected to the staff of Lecturers in Architecture, and will, during the coming year, give a course on the History of Architecture.

Mr. F. M. Mann retired from the position of Instructor in Architecture. Mr. Thomas Nolan, who was formerly an Instructor in the department of Architecture, has been appointed to his old position, and Mr. C. F. Osborne has been elected Instructor in the History of Architecture.

The department of European History lost a valued instructor by the resignation of Dr. Merrick Whitcomb, who was elected Professor of History in the University of Cincinnati. Dr. Whitcomb left us followed by the good wishes of all his colleagues, who look with pleasure upon the deserved promotion and recognition which his new position carries with it. Dr. Whitcomb's classes will be taken during the coming year by Mr. W. E. Lingelbach, who has been appointed Instructor in European History.

The department of French suffered loss in the death of Dr. Samuel P. Molenaer, who died in June, after several months of severe illness. The work that he did is remembered and appreciated by his colleagues, and by the students who came under his instruction, while his faithful and earnest service of the University has been a matter of general recognition. He was the leader of the movement to establish, among the undergraduates of the College, a French Club; and under his efficient direction this club produced, most creditably, on the evenings of March 29 and 30, the French play, "Le Medecin Malgrè Lui." The work of drilling the students who took part in the play was among the last duties performed by Dr. Molenaer, who was, however, too ill to share in the success of the public performance. During Dr. Molenaer's illness, his courses were given by Mr. Vurpillot, who has been elected his successor.

The tragic death of Mr. Roy Wilson White was a serious loss to the College, in which he gave courses on

Roman Law. Extended mention of the services of Mr. White belongs more properly to the report of the Dean of the Department of Law, of the staff of which Mr. White was a member.

Dr. J. B. Moyer, who for seven years was Instructor in Chemistry, resigned to accept the position of Professor of Chemistry in the Philadelphia College of Dental Surgery. Dr. W. L. Hardin retired from his position as Instructor in Technical Chemistry to enter technical pursuits. Mr. J. Gillinder, Jr., has been appointed Instructor in Chemistry.

As noted in the last report, the duties of Assistant Professor E. R. Johnson, as a member of the Isthmian Canal Commission, kept him from his University duties during the year, and rendered necessary changes in the arrangement of the courses given to students in Finance and Economy. Dr. H. S. Pratt, of Haverford College, was appointed Lecturer on Commercial Geography, and conducted some of the courses of Assistant Professor Johnson.

At the end of the year two other members of the teaching force of the College were honored by appointments to government positions by the President, and were granted leave of absence by the University. Professor Brumbaugh was given the important position of Commissioner of Education for Porto Rico. The Graduate and Teachers' courses of Professor Brumbaugh will, during his absence, be conducted by Dr. N. C. Scheaffer, State Superintendent of Public Instruction, while the undergraduate courses will be in charge of Mr. J. C. Hockenberry.

Assistant Professor L. S. Rowe was appointed a member of the commission in whose charge was placed by Congress the task of revising and compiling the laws of Porto Rico.

Public service claimed yet another of our men. Associate Professor Roland P. Falkner, who for twelve years was connected with the department of Finance and Economy, resigned his University position to accept the newly-created office of Chief of the Bureau of Public Documents, in the Congressional Library at Washington. Dr. Falkner takes with him the esteem of his former associates and students; and the public service will profit by the appointment of a man so eminently fitted for the new duties he has now undertaken. It is with regret that the University parted with Dr. Falkner.

The absence of Assistant Professors Johnson and Rowe, and the establishment of a new four-year course in Commerce and Industry, led to the appointment of new instructors in the courses in Finance and Economy, with the following titles: Dr. E. S. Meade, Instructor in Industry and Banking; Dr. F. A. Cleveland, Instructor in Finance; Dr. W. H. Allen, Instructor in Public Law.

Mr. A. R. Bauder, Instructor in Physics; Mr. G. M. Richter, Assistant in German, and Mr. H. D. Tiemann, Instructor in Mechanical Engineering, will be succeeded by Messrs. W. McClellan, E. C. Parry and S. M. Marshall, respectively.

New System of Faculty Advisers for Students.

The Academic Council adopted a new plan for keeping students' records, and at the same time provided what has been needed for some time—an efficient system of advisers for students. Instead of relying, as heretofore, on the catalogue announcements for the correct statement of the work to be done by each student in a given course, it was decided to make the students and their respective advisers responsible for the filing in the office of cards containing such statements. Under the new system the students have been divided into small groups, for each of which a member of the teaching staff has been appointed adviser. Students are referred to their advisers in regard to all matters. In this way the students are brought into closer personal relations with the Faculty.

The cards filed in the office by the students, after consultation with their advisers, will serve as current records during the term, and at the end of the term will be issued with the grades or marks as reports. The permanent records of the College will be made from these cards and entered on the books.

Change in the System of Collecting Fees.

By action of the Board of Trustees, the system of collecting fees in the various departments of the University was made uniform. No bills are sent out, but all students are informed that their tuition is due in two instalments, payable respectively on October 1 and February 1; and that a penalty of two dollars and a half will be added to all bills not paid before November 1 and March 1 thereafter. The results of the new system have been entirely satisfactory in the College.

New Arrangement Between the College and the Department of Medicine.

In the last annual report mention was made of the establishment of closer relations between the College and the Department of Law, by providing in the Senior and Junior years in the courses in Arts and Science a new group of studies in Economics, Social Science and Law.

The College and the Department of Medicine have been brought together more closely by a number of changes which affect the upper years in the courses in Arts and Science, and the second year of the special course in Biology. The provisions by which college graduates possessing certain qualifications in Physics, Chemistry and Biology were formerly admitted to the second year in Medicine have been abolished. Hereafter all graduates in Medicine from this University must have been registered for four full years as students in the Department of Medicine, pursuing the regular Medical course. The General Chemistry and the Histology and Embryology of the first year in Medicine, however, may be done in the regular classes in these subjects in the College; and this work, when certified to by the proper College authorities, will hereafter be accepted by the Department of Medicine in lieu of their first-year work in these subjects. The other subjects of the first year in Medicine, viz., Anatomy, Physiology, Bacteriology and Physiological Chemistry, must in all cases be done in the Medical Department; and the College will hereafter accept any or all of these subjects as a part of the regular work of the Senior year in Group XIV of the Arts and Science courses, and also in the second year of the two-year course in Biology.

In making these provisions the University has been guided by two considerations: First, the unification of the work of the University; second, the recognition by one University department of equivalent work done in another department. (1) The work of the first year in Medicine as now arranged is work in pure science, and as such is entitled to a place in the College curriculum. In fact, all the subjects of the first year in Medicine, viz., Anatomy, Physiology, Bacteriology, Histology and Chemistry, have until recently been given in the College as well as in the Department of Medicine: but in order to avoid needless duplication of courses, and to provide the best possible facilities for work in those subjects, certain of these courses have been unified, and College students who desire to study Human Anatomy, Human Physiology or Bacteriology are now permitted to take these subjects in the Department of Medicine; whereas medical students who desire work in subjects offered by the College have a similar privilege extended to them. (2) The Department of Medicine has for some time accepted the work in General and Analytical Chemistry done in the College in lieu of the first-year Chemistry of the course in Medicine; and under certain restrictions similar recognition has been accorded the work of the College in Histology and Physiology. The College, on the other hand, has not heretofore accepted any of the work of the Department of Medicine in lieu of its own work. The present provisions look to the recognition by either department of certain work which has been done in one department, and which may belong to both.

These new provisions affect particularly Group XIV of the Arts and Science courses, and the two-year course in Biology. The former will hereafter be known as the Botany-Zoölogy-Medicine Group.

Students who elect this group, and are at the same time matriculated in both the College and the Department of Medicine, may receive the Bachelor's degree with their College class at the end of the Senior year, and at the same time may obtain full credit for the first year's work in Medicine; so that the degrees of B. S. or A. B. and M. D. may be taken in seven years.

The two-year course in Biology has been modified as follows: The work of the first year is the same for all students. In the second year two alternative courses are open: the first is a continuation of the work of the first year in the Natural Sciences; and the second consists of the first year's work in Medicine plus certain subjects given in the College. Human Anatomy, Human Physiology, Bacteriology and Physiological Chemistry are taken in the Department of Medicine; Histology and Embryology, and Chemistry in the College.

Students who choose the second alternative, and are matriculated as regular first-year students in the Department of Medicine, are, on passing the required examinations, advanced with their class to the second year in Medicine. The fee of all students, whether in Group XIV or in the two-year course, who are matriculated in the Department of Medicine, is $200 per annum. Men only are admitted to courses in the Department of Medicine.

New Four-Year Course in Commerce and Industry.

The two-year special course in Commerce, Diplomacy and International Law was abolished by action of the Academic Council, as were also the two- and three-year courses in Finance and Economy. A new four-year course was established with the title *The Course in Commerce and Industry.* It leads to the degree of Bachelor of Science in Economics, and the requirements for admission are the same as those for the course in Finance and Economy. This new course is designed to meet the need of commercial education of collegiate grade which has been keenly felt in business circles.

The course recognizes that business knowledge is a requisite to success as well as business habits, and seeks to impart general and technical information which will be helpful. While its instruction cannot be a substitute for business experience, it offers for the future man of affairs a training analogous to that which faculties of law and medicine offer to men preparing for those professions.

In addition to giving the student the discipline of a college education, the course aims to furnish him with a special knowledge of mercantile and industrial activities, which will fit him to take a place in the higher walks of commercial life, whether in internal or foreign trade, in the conduct of manufacturing or transportation enterprises, or in the consular service. The basis of instruction lies in economic and political science, to which is added the study of business law, business methods as exemplified especially in banks, railroads and shipping, the history and present condition of both trade and industry. The course also includes modern languages, which are of increasing importance in the commercial world.

Lengthening of the Course in Music.

The course in Music has been lengthened from three years to four; and, in addition to a knowledge of the rudiments of music and ability to play on some musical instrument, candidates for admission to the course will be required to pass the College entrance examination in English. The requirements for the degree of Bachelor of Music have been modified to accord with the advanced requirements for the certificate given on completion of the four-year course, and in addition to the work represented by the certificate; candidates for the degree must pursue at least three of the five courses in English Language and Literature offered for students in the Saturday Classes for Teachers. The technical requirements for the degree remain the same as heretofore.

Additional Requirements for Special Course in Architecture.

The entrance requirements for admission to the special courses in Architecture have been increased by the addition of Algebra, Plane and Solid Geometry, and Freehand Drawing to the two years' experience in an architect's of-office or its equivalent, which has always been required.

Additions to Equipment.

The department of Physics received from Mr. George Westinghouse, of Pittsburg, through the kind agency of Dr. Coleman Sellers, a ten-horsepower rotary transformer, with its appropriate exciting dynamo of about one horse-power. The machine is adapted to run as a motor, with a direct current of 110 volts, and to deliver a two-phase alternating current. It may be used either for practical purposes or for demonstration.

Another interesting piece of apparatus added to the equipment of this department is a specially-constructed instrument for demonstrating the presence of, and the properties of, a rotating magnetic field of great strength. It is admirably operated by the rotary transformer mentioned above; and was also a gift, with the compliments of Mr. Westinghouse. The apparatus was constructed for Dr. Sellers, for the purpose of demonstrating the principle of the rotating field to Mr. Maskelyne, of London, who visited this country in the spring to view and photograph the eclipse of the sun.

Mention may properly be made here of the high quality and permanent value of the thesis work done by students in the course in Biology; and this work has been greatly facilitated by such additions as have been made to the equipment of the courses in Botany and Zoölogy, both in books and apparatus. At the beginning of the year two large paraffine ovens were provided, furnishing individual accommodations for thirty-two students. A large number of anatomical jars were purchased; and through the generosity of

Nicholas Thouron, Esq., of the Class of 1864, four large microscopes were obtained for the use of advanced students in Zoölogy.

A new course in General Zoölogy, dealing with the fundamental properties of animals, their general anatomy, physiology, development, ecology and evolution, was given during the year with very satisfactory results.

During the year a complete outfit for electrolytic work in Inorganic and Organic Chemistry has been introduced. Generous friends, to whom the Provost presented our needs, gave the money necessary for this purpose. As it now stands, we have the most recent and most complete laboratory of this kind in this country.

The additions to the equipment of the department of Geology and Mineralogy during the past year were as follows:

Two hundred lantern slides, prepared in the Laboratory for illustration of lectures; 400 Mineral and Ore specimens, purchased; 500 Mineralogical and Geological specimens, presented by Professor Brown; 350 Mineralogical and Geological specimens, presented by Dr. F. Ehrenfeld; a very fine specimen of lead ore from Joplin, Mo.; a unique specimen, presented by Mr. Rafael Estrada, through Professor Houston; about 25 miscellaneous specimens from various sources.

To the equipment of the department of Mechanical and Electrical Engineering were added during the year a jet condenser, with its air-pump, one universal milling machine, two steam-engine indicators, a gas testing apparatus, two planimeters, two water meters, voltmeters and ammeters, telescope, two resistance frames and two condensers.

The Need of a New Building for the Departments of Engineering.

Attention may be called to the crowded condition of the Engineering departments, especially in the matter of laboratory space and drawing room. In the Mechanical

Drawing-Room there are fifty-three desks, which is really a greater number than should be put in the space; and at times this room is occupied by sixty-two men. In the Laboratory the space is so contracted that it is often necessary to dismantle one piece of apparatus before work can be carried on with another.

The obvious solution of this problem is a new Engineering building. It requires no argument to show that Philadelphia is in every way peculiarly fitted to become a great centre of engineering education.

Annual Convention of City and Borough Superintendents of Schools.

On February 8, 9 and 10, 1900, the University had as its guests the City and Borough Superintendents of Schools of Pennsylvania; and the deliberations of the annual meeting of this body was concerned largely with the questions affecting the correlation of school and college courses. This problem, one of the most important in our whole educational system, will find its solution only when the schools and colleges are brought into close association in all deliberations affecting the interests of both.

Conference on College Entrance Examinations.

The conference held at Columbia University two years ago, for the purpose of securing greater uniformity in the interpretation of college entrance requirements, and for the establishing of a joint body of examiners composed of representatives of fitting schools as well as of colleges, seems to have met with general approval. Last year two meetings were held for the purpose of arranging the necessary details.

Student Organizations.

To the number of Student Organizations already existing, there was added last year a Woman's Club, composed of the women students in the College and Departments of Philosophy and Law. In addition to the Philomathean and Zelosophic Societies of the College, and the Debating Union, in which the students of the College are taking an active interest, there exist a number of organizations among the students pursuing the various courses, the objects of these being both social and scientific. Such clubs exist in connection with the work in Botany, Zoölogy, Architecture, Civil Engineering, Mechanical and Electrical Engineering, Chemistry and Music.

The Cercle Francais.

During the fall the University of Pennsylvania and Haverford, Bryn Mawr and Swarthmore Colleges united in the organization of a Cercle Français, the meetings of which were well attended.

Lectures and addresses were delivered in the Chapel and in the Auditorium of Houston Hall as follows:

M. Victor Rigueur, "Victor Hugo—Quatre-Vingt-Trieze."

M. Alphonse de Chateaneuf, "Le Parisien, Heretier de Son Oncle de Californie." (Reading.)

Mr. H. V. Le Maistre, of New York, "Le Voyage de Monsieur Perrichon." (Reading.)

Professor Adolph Cohn, of Columbia University, "L'Academie Française en Dix-Neuf."

Professor Albert Schinz, Ph. D., of Bryn Mawr College, "Le Groupe des Poetes de Parnasse."

Mme. Marion, "Paris et ses Monuments." (Illustrated.)

M. Henri de Regnier, of Paris, "The Poets of To-day." Works and Biography. "Alfred de Vigny." "Women in Contemporary Literature."

Addresses in Chapel.

The one hundredth anniversary of the death of Washington was appropriately marked by special exercises in the Chapel on December 14, 1899, at which Judge Pennypacker addressed the students. Other addresses in Chapel were delivered during the year by the following speakers:

Rev. Edward Everett Hale, of Boston.
Canon Edmonds, of Exeter Cathedral, England.
Hon. James M. Beck, U. S. District Attorney.
Samuel F. Houston, Esq., of the Board of Trustees.
Rev. Dr. Floyd Tomkins, of Philadelphia.
Dean Hart, of Denver, Colorado.
Dr. Rendel Harris, of Cambridge.
Rev. Dr. Stubbs, Dean of Ely Cathedral, England.

University Committee on Students' Residences.

By action of the Board of Deans, the Committee on Students' Residences, which has been in existence for some years in the College, was enlarged and made a University Committee, with the same duties as heretofore. Inquiries are made of students in regard to their boarding houses, and all boarding houses at which students live are visited by members of the Committee. The sanitary condition of the houses, the quality and price of the meals supplied, and the various advantages and disadvantages of the houses are recorded for the benefit of students who may desire to secure accommodation. A printed list of recommended boarding houses may be obtained by students on application at the office of the Deans of their respective Departments.

Gifts of Classes.

The Class of 1872 has presented to the University a handsome memorial gateway, which has been erected at the east end of Hamilton's Walk.

The Class of 1894 presented the iron gates which have been placed at the entrance to the Memorial Tower.

The Need of Endowment.

In closing this report, attention must again be called to the need of adequate endowment for the College. A glance at the report of the Treasurer will show how great that need is, for the annual deficit in the College is greater than the total annual expenses of many institutions. The time will never come when the income from tuition alone will make the College self-sustaining, for no student in a College of grade similar to ours can be charged for tuition a sum equal to his share of the cost of providing the courses of instruction. It follows necessarily that the deficit must be met either by gifts from time to time to meet current expenses, or by permanent and sufficient endowment. At present rates, two million dollars are needed to lift from the shoulders of the Provost and Trustees an annual burden that is hard to bear.

Respectfully submitted,

JOSIAH H. PENNIMAN,

Dean.

COLLEGE STATISTICS FOR 1899–1900.

TABLE I.

STUDENTS IN COLLEGE, 1899–1900.

					Sc., Science and Technology.				Four-Year Technical Courses.							
	A., Arts and Science.	Wh., Finance and Economy.	B., Biology.	M., Music.	Sc., Specific Courses not Decided.	Sc. 1, Chemistry.	Sc. 2, Civil Engineering.	Sc. 3, Mechanical Engineering.	Ch., Chemistry.	Ch. E., Chemical Engineering.	C. E., Civil Engineering.	M. E. and H. E., Mechanical and Electrical Engineering.	Arch., Architecture.	I. Dec., Interior Decoration.	Totals.	
Post-Seniors	...	...	...	...	...	1	2	2	...	...	...	...	...	...	5	Post-Seniors.
Seniors	35	9	4	...	...	1	1	4	10	1	5	17	7	...	94	Seniors.
Juniors	52	22	3	...	...	...	...	6	8	4	7	27	8	...	137	Juniors.
Sophomores	43	23	10	...	3	...	...	...	9	4	15	12	8	...	127	Sophomores.
Freshmen	40	25	4	...	5	...	...	...	8	2	16	30	15	...	145	Freshmen.
Students in Special Courses: Third Year	...	1	...	17	...	1	...	...	...	...	...	...	...	3	22	Third Year. (Students in Special Courses.)
Second Year	...	8	1	10	...	2	...	...	...	...	...	...	7	2	30	Second Year.
First Year	...	19	5	15	...	...	...	...	...	...	...	...	8	4	51	First Year.
Partial Students	8	18	28	...	...	3	...	1	17	...	11	5	3	1	95	Partial Students.
Totals	178	125	55	42	8	8	3	13	52	11	54	91	56	10	706	Totals in the College.

Students taking Courses for Teachers . . . 262

Total 968

TABLE II.

SHOWING ADMISSIONS TO COLLEGE, 1899.

Average age of students entering Freshman class in '98, 18 years, 3 months, 14 days.

Average age of students entering Freshman class in '99, 18 years, 6 months, 11 days.

Total number of applicants		376
Applicants admitted on examination	82	
Applicants admitted on public high school diploma	83	
Applicants admitted on normal school diploma	4	
Applicants admitted on credits from other colleges	15	
Applicants admitted to advanced standing	13	
Applicants admitted to special and partial courses	101	
		298
Applicants rejected on examination		31
Applicants taking preliminaries and receiving credits	41	
Applicants taking preliminaries and receiving no credits	6	
		47
Total		376

APPLICATIONS CONSIDERED BY THE COMMITTEE ON ADMISSION TO SPECIAL AND PARTIAL COURSES.

Applications on which Committee took action		110
Total applications approved	101	
Total applications rejected	9	
		110

APPLICATIONS CONSIDERED BY THE COMMITTEE ON ADMISSION TO ADVANCED STANDING.

Applicants who entered college	28	
Applicants who did not enter college	20	
		48

APPLICATIONS CONSIDERED BY THE COMMITTEE ON SCHOOLS AND ENTRANCE REQUIREMENTS.

Total number of cases acted upon		87
Applicants admitted without examination on diploma	39	
Applicants admitted on diploma with supplementary examination	35	
Applicants rejected	1	
		75
Cases considered where applicants did not enter college		12
Of these, admitted without condition	6	
admitted with conditions	6	
Total		87

CONDITIONS IMPOSED BY COMMITTEE ON SCHOOLS AND ENTRANCE REQUIREMENTS ON CANDIDATES PRESENTING DIPLOMA OF A PUBLIC HIGH SCHOOL.

English B	1
History B	1
History D	1
History C	1
Mathematics C	5
Mathematics E	3
Mathematics F	6
Latin A	2
Latin B	2
Latin C	6
Latin D	5
Latin E	5
Latin F	9
Latin G	4
French A	2
French B	2
German A	16
German B	18
Physics	1
	90

TABLE III.

SHOWING ELECTIONS OF GROUPS MADE BY JUNIORS AND SENIORS IN ARTS AND SCIENCE.

	1899–1900.	
	Seniors.	Juniors.
Greek-Latin	9	8
Latin-German	2	2
French-German	1	5
English-German	8	4
English-French	2	4
History-English	2	5
Philosophy-History	4	2
History, Politics and Economics	6	9
Physics-Mathematics	1	1
Chemistry-Physics	1	—
Biology-Chemistry	1	1
Botany-Zoölogy	—	1
Geology-Chemistry	—	1
Economics, Social Science and Law	—	7
Total	37	50

TABLE IV.

(1) NUMBER OF STUDENTS CONDITIONED AND NUMBER OF CONDITIONS IMPOSED.

	Number of Students Conditioned.	Number of Conditions.	Conditions per Student Conditioned.
FIRST TERM.			
Post-Seniors	—	—	—
Seniors	9	9	1.
Juniors	46	81	1.8
Sophomores	47	67	1.4
Freshmen	69	172	2.5
Total Regulars	171	329	1.9
Specials, third year	2	3	1.5
Specials, second year	10	18	1.8
Specials, first year	17	38	2.2
Total Specials	29	59	2.0
Partials	34	90	2.6
Grand Totals	234	478	2.0
SECOND TERM.			
Post-Seniors	—	—	—
Seniors	5	7	1.4
Juniors	29	45	1.6
Sophomores	31	53	1.7
Freshmen	49	90	1.8
Total Regulars	114	195	1.7
Specials, third year	—	—	—
Specials, second year	5	7	1.4
Specials, first year	8	12	1.5
Total Specials	13	19	1.4
Partials	22	38	1.7
Grand Totals	149	252	1.7

TABLE V.

(2) NUMBER OF STUDENTS ADVANCED WITH CONDITIONS AND NUMBER OF CONDITIONS CARRIED.

	Number of Students Allowed to Continue with Conditions.	Number of Conditions Carried by these Students.	Number of Conditions per Student Conditioned.
FIRST TERM.			
Post-Seniors	—	—	—
Seniors	23	44	1.9
Juniors	58	111	1.9
Sophomores	58	106	1.8
Freshmen	41	107	2.6
Total Regulars	180	368	2.0
Specials, third year	1	2	2.0
Specials, second year	6	15	2.5
Specials, first year	17	39	2.3
Total Specials	24	56	2.3
Partials	32	76	2.4
Grand Totals	236	500	2.1
SECOND TERM.			
Post-Seniors	—	—	—
Seniors	5	6	1.2
Juniors	43	94	2.2
Sophomores	37	75	2.0
Freshmen	40	93	2.3
Total Regulars	125	268	2.1
Specials, third year	—	—	—
Specials, second year	4	11	2.7
Specials, first year	14	35	2.5
Total Specials	18	46	2.6
Partials	29	75	2.6
Grand Totals	172	389	2.3

TABLE VI.

COLLEGE STUDENTS BELONGING TO UNIVERSITY ATHLETIC TEAMS.

(Football, Baseball, Crew, Track, Gymnasium and Cricket.)

NUMBER OF STUDENTS ON THESE TEAMS:

Regulars	87
Specials	7
Partials	12
Total	106

	None Dropped.	None Dismissed.	Withdrawn. 1st Term.	Withdrawn. 2d Term.
Regulars			—	4
Specials			—	—
Partials			1	3

	Number of Students Conditioned.	Number of Conditions.	Number of Conditions per Student Conditioned.
FIRST TERM.			
Regulars	34	57	1.7
Specials	5	7	1.4
Partials	11	20	1.8
Totals	50	84	1.7
SECOND TERM.			
Regulars	31	54	1.7
Specials	3	5	1.6
Partials	7	14	2.0
Totals	41	73	1.8

TABLE VII.

STUDENTS BELONGING TO ORGANIZATIONS OTHER THAN ATHLETIC (MASK AND WIG, UNIVERSITY BAND, THE RED AND BLUE, AND PENNSYLVANIAN).

NUMBER OF STUDENTS IN THESE ORGANIZATIONS:

Regulars	107
Specials	16
Partials	9
Total	132

	Dropped. 1st Term.	Dropped. 2d Term.	Withdrawn. 1st Term.	Withdrawn. 2d Term.
Regulars	2	—	3	4
Specials	1	—	—	—
Partials	1	—	1	—
Totals	4	—	4	4

	Number of Students Conditioned.	Number of Conditions.	Number of Conditions per Student Conditioned.
FIRST TERM.			
Regulars	38	54	1.2
Specials	6	13	2.2
Partials	5	12	2.4
Totals	49	79	1.6
SECOND TERM.			
Regulars	17	24	1.4
Specials	2	3	1.5
Partials	6	7	1.2
Totals	25	34	1.4

TABLE VIII.

STUDENTS WITHDRAWN OR DROPPED.

	Withdrawn.	Dropped or Put Back. 1st Term.	2d Term.	Dismissed or Suspended	Deceased.
Post-Seniors	—	—	—	—	—
Seniors	1	—	—	—	—
Juniors	5	2	—	2	—
Sophomores	7	—	—	—	—
Freshmen	6	4	3	—	1
Total Regulars	19	6	3	2	1
Special, third year	2	—	—	—	—
Special, second year	2	1	—	—	—
Special, first year	2	3	—	—	—
Total Specials	6	4	—	—	—
Partials	20	6	2	—	—
Grand Total	45	16	5	2	1

TABLE IX.

SCHOLARSHIPS.

	Arts.	Wharton.	Science.	Mechanical Engineering.	Civil Engineering.	Chemistry.	Biology.	Architecture.	Chemical Engineering.	Grand Total.
Full Faculty Scholarships.										
Post-Seniors	—	—	3	—	—	—	—	—	—	3
Seniors	10	—	3	4	—	4	1	2	—	24
Juniors	6	2	—	2	1	3	2	1	—	17
Sophomores	2	1	—	1	3	1	—	2	—	10
Freshmen	5	2	2	—	—	1	—	1	—	11
Partial	—	1	—	—	—	—	—	—	—	1
Total	.	.	.	.	.	.	.	.	.	66
Partial Faculty Scholarships.										
Seniors	1	—	—	—	—	1	1	—	—	3
Juniors	4	1	1	3	1	1	—	—	1	12
Sophomores	—	—	—	—	—	—	1	—	—	1
Freshmen	—	—	—	—	—	—	—	—	—	—
First Special	—	—	—	—	—	—	—	—	—	—
Second Special	—	—	—	—	—	—	—	—	—	—
Total	.	.	.	.	.		.	.	.	16
Muhr Scholarships.										
Senior	1	—	—	—	—	—	—	—	—	1
Junior	—	—	—	—	1	—	—	—	—	1
Sophomore	1	—	—	—	—	—	1	—	—	2
Total	.	.	.	.	.	.	.	.	.	4
Penn Scholarships.										
Junior	—	—	—	—	—	1	—	—	—	1
Freshmen	1	—	—	—	—	—	—	—	—	1
Total	.	.	.	.	.	.	.	.	.	2
Full Free City Scholarships.										
Senior	5	1	—	1	2	—	—	—	—	9
Junior	4	—	—	2	1	—	1	—	—	8
Sophomore	2	1	—	—	—	—	1	—	1	5
Freshmen	3	—	—	3	3	2	2	1	—	14
Total	.	.	.	.	.	.	.	.	.	36
Partial Free City Scholarships.										
Sophomore	2	3	—	1	4	1	6	—	1	18
Freshmen	1	—	—	3	3	2	3	—	—	12
First-Year Special	—	—	—	—	—	—	2	—	—	2
Total		.	.	.	.	.	.	.	.	32
Totals	4[illegible]	12	9	20	19	17	21	7	3	156

APPENDIX III.

REPORT OF THE DEAN OF THE DEPARTMENT OF PHILOSOPHY.

To the Provost of the University;

SIR:—I have the honor of presenting to you my report upon the progress of our work during the year ending August 31, 1900, as follows:

THE FACULTY AND ITS COMMITTEES.

The members of the Executive Committee for the year 1899-1900 were:

Edgar F. Smith, Vice-Provost, *ex-officio,*
William Romaine Newbold, Dean, *ex-officio,*
William A. Lamberton,
Simon N. Patten,
George Stuart Fullerton,
John M. Macfarlane,
Marion D. Learned.

Professor A. W. Goodspeed has represented this Faculty upon the University Athletic Committee, and Professor E. P. Cheyney upon the University Committee on Non-Athletic Organizations.

The Committee on Discipline has consisted of the Dean, Professors Crawley and Conklin. It has had no duties to discharge.

Professor Hilprecht having been absent during the past year in the service of the University, Dr. Clay has given, with credit to himself and to the University, many of the courses of instruction offered by Professor Hilprecht.

Assistant Professor Emory R. Johnson has also been absent on leave throughout the year. No provision having been made for his work, it has been allowed to lapse for the present.

Professor Brumbaugh was granted leave of absence towards the close of the year to accept the position of Commissioner of Education for the Island of Porto Rico. During the coming year his graduate work will be carried on by Dr. Nathan C. Schaeffer, Superintendent of Public Instruction for the State of Pennsylvania. The roster has been so arranged that Dr. Schaeffer's presence in Philadelphia will be required for not more than twenty hours in the week, thus avoiding conflict with his important duties to the State.

Assistant Professor Rowe was also granted leave of absence at the close of the year to accept a position upon the Commission appointed by the President of the United States for the codification of the laws of Porto Rico. I regret to say that it has not been found possible to make provision for his work in Political Science.

We have lost by resignation, Dr. Merrick Whitcomb, Lecturer in European History; and Dr. Roland P. Falkner, Associate Professor of Statistics. The services which Professor Falkner has rendered to this Department for the past twelve years, and Dr. Whitcomb for three, have been highly appreciated by both students and colleagues.

The close of this year witnessed also the retirement of one of the University's oldest, most faithful and most distinguished professors from the scene of his active labors. George F. Barker, M. D., Sc. D., LL. D., one of the few surviving founders of this Department, and Professor of Physics in it since its foundation, was compelled by failing health to give up his work. Professor Barker's devotion to his duties, his brilliant success as a lecturer, his eminence in his profession, and the cordial personal relations which he maintained with his colleagues, are well known to all the alumni and friends of the University. But of his personal relations with the administrative officers, and especially with those who belonged to a younger generation than himself, I feel that a word of appreciation will not be out of place. He has always shown himself most cordial and most helpful—as willing to aid in carrying out plans laid in the first instance by others as those in which he had a more personal interest. Our regret in losing his counsel and assistance is, however, tempered by the reflection that the well-earned rest to which he now retires will doubtless prolong his life, and will not wholly remove his personal influence from the University.

Dr. Henry W. F. Lorenz has been appointed Lecturer in Chemistry.

The position left vacant by the death of Professor D. G. Brinton has been filled in part by the appointment of Mr. Stewart Culin to the position of Lecturer in Ethnology and American Archæology. Mr. Culin had been for many years associated closely with Professor Brinton, and has served the University in its Museum and in directing many of its archæological expeditions. His contributions to the history of culture, more especially in connection with the origin and primitive significance of Games, have been well received, both in this country and abroad. He will offer work in American Archæology analogous to that offered by Professor Brinton. He will also offer work in Ethnology not given by Professor Brinton, but will not give the instruction in American Languages which was a leading feature of Professor Brinton's work.

In pursuance of these modifications of Group Committee No. II's work, the work in Archæology offered by other Group Committees has been transferred to its jurisdiction, and several new courses have been planned. Professor Hilprecht's title was changed from "Professor of Assyrian and Comparative Semitic Philology" to "Professor of Semitic Philology and Archæology." Dr. Clay's title was changed from "Lecturer in Assyrian and Hebrew" to "Lecturer in Assyrian, Hebrew and Semitic Archæology." Dr. Bates's title was changed from "Lecturer in Greek" to "Lecturer in Greek and in Classical Archæology." Professor Hilprecht, Dr. Clay and Dr. Bates, together with Mr. Culin, were made members of Group Committee II; and the title of that Committee was changed from "American

Languages and Archæology" to "Ethnology and Archæology." In accordance with our general provisions, Group Committee II has arranged a series of courses for the coming year. Professor Hilprecht has planned two standard courses, one in Babylonian Paleography, and one in Life and Customs of the Early Babylonians. Dr. Clay has offered one hour in Hebrew Archæology; and Dr. Bates has offered two hours—one in Greek Epigraphy and one in Greek Inscriptions.

In the course of the year, the Faculty of the Department of Law offered to give under the jurisdiction of this Faculty eighteen hours of instruction in Comparative Jurisprudence, the History of Jurisprudence, and problems of the borderland between Law and Economics. The work was to be given by Mr. Roy Wilson White, Assistant Professor W. E. Mikell and Dean William Draper Lewis, and was to be arranged in a three-year cycle, six hours a year, beginning October 1, 1900. The same work was also to be offered by the Faculty of Law to candidates for the degree of Master of Laws. This offer was accepted by the Faculty of Philosophy; the legislation necessary to carry it into effect was adopted; Professor Lewis and Assistant Professor W. E. Mikell were put upon the Faculty of this Department; and Mr. White was authorized to give instruction. This action of the Faculty was approved by the Board of Trustees.

The sudden death of Mr. White, May 19, 1900, has, however, made it impossible to carry out these plans during the coming year. I have every reason to believe that this additional work will be offered during the year 1901-02.

Since this work is designed to train students, not in the practice of law, but in the social and economic conceptions which underlie all legal codes; and to trace the modifications in those conceptions, and the consequent modifications in current laws which recent economic developments have occasioned; it falls properly within the scope of this Faculty's aims, and would long since have been offered by it, had our resources permitted. I think, therefore, that the action of the Faculty of Law, from whom in the first instance the suggestion came, can claim the appreciation of all who are interested in the development of our work in those higher branches of learning in which the love of knowledge for its own sake is the ruling principle. It is also a further and gratifying indication of that increase in the sense of community of interest between the several departments of the University, and in the spirit of friendly co-operation between them, which has been so noteworthy a feature of the more recent years of our history.

Administration.

Important changes have been made in the regulations under which candidates are allowed to proceed to the Master's and Doctor's degrees.

From the organization of this Department in 1882 to the end of the year 1895-96, the final examinations required of candidates were oral. These examinations were conducted by a Committee consisting of the Dean, and three members of the Faculty chosen by him, representing the candidate's three subjects. The examinations were public. All members of the Faculty were invited to attend, and were at liberty to question the candidate if they saw fit to do so. Very few of the members of the Faculty have at any time taken advantage of this invitation, excepting those under whose instruction the candidate had come. The consequence was that the final examinations were, in most cases, conducted by men who were already thoroughly familiar with the candidate's merits, and whose judgment was not likely to be materially modified by the results of two hours' oral examination. In order to meet this difficulty, and to provide the examiners with more substantial and visible grounds for their judgment, a regulation was adopted in May, 1896. This required that all candidates for the higher degrees should pass written examinations before being admitted to the oral examinations; and that their written answers should be presented to the Committee at the oral examination, and should form part of the material upon which the final decision was rendered. The immediate effect of this provision was a further depreciation in the value attached to the oral examination. The written examinations were much more prolonged than the oral, and proportionately more searching. The candidate who presented himself for a two-hour oral examination had already passed written examinations aggregating from fifteen to thirty hours; and where he failed to do as well in the oral as he had done in the written, as sometimes happened, the decision of the Committee was apt to be influenced by the results of the written rather than by those of the oral examination.

At the opening of the year 1899-1900 the matter was brought up for discussion. The value of the written examinations had been thoroughly tested, and no question as to their continuance was raised. They will in future be made even more searching. But as regards the oral examination, an overwhelming majority of the Faculty was convinced that as then conducted it no longer subserved any useful purpose. Two methods of reform were suggested. The first was a return to the earlier practice, still retained in many institutions, of bringing the candidate before the entire Faculty and compelling him to satisfy them, as well as the men with whom his work had been done, that he was worthy of the degree he sought. The other was the abolition of the oral examination.

The first step recommended itself to many minds, not only because it was sanctioned by ancient practice, but because it brought the full light of publicity to bear upon the candidate's claims. Nothing tends more directly to safeguard the higher degrees, and to prevent the development of diversities of standard within an institution, or between one institution and others, than regulations insuring publicity.

After careful study of the situation, however, we came somewhat reluctantly to the conclusion that such a return to earlier methods would be, under our present conditions, impracticable; and even if practicable, unadvisable. The rapid growth of specialization among the members of the teaching staff in all modern universities has rendered the greater number unfit to judge of the merits of a candidate who has specialized in any branch of study far removed from their own branches. However we might wish it were otherwise, the final decision must be rendered by those who are working in lines akin to those of the candidate; and to their decision, after removing so far as possible all agencies that might affect it unfavorably, we must trust. Other objections to such a step were advanced, among them being the fact that a series of more than twenty-five two-hour examinations, requiring the attendance of the entire Faculty in the latter part of May, would disorganize the work of the College for at least a week at the busiest time of year; and the further fact that a strong sentiment exists in this Faculty against a step to which no advantages were attached that could not be had otherwise. But as these were of merely local significance, I feel that I need give them but cursory mention.

Against the abolition of the oral examination, without providing anything to take its place, one chief objection was urged. Quite apart from the value of the oral examination as a test of the candidate's fitness, it has been regarded as a suitable method of bringing the candidate's course of study to a conclusion, and of formally expressing the Faculty's approval of it and of him. This feature we were all unwilling to see obliterated from our practice.

The regulations finally adopted will, it is believed, preserve what is best in the traditional formalities, while setting aside what has been found unsuited to the conditions of the present day. They are based upon two fundamental conceptions:

1. The recognition of the fact that the candidate's fitness for the Ph. D. degree must be determined by those members of the Faculty who are themselves working in, or not far from, the lines in which he has specialized.

2. The making of those members of the Faculty responsible to public opinion by insuring for their recommendation, and for the grounds upon which it is based, the greatest possible degree of publicity, both within and without the institution.

To insure this publicity within the institution, and to mark the conclusion of the student's work with suitable ceremony, it has been provided that a formal meeting of the Faculty shall be held, the Provost being in the chair, at which meeting every candidate for the Master's and Doctor's degrees shall appear in person. The candidates for the Master's degree will be presented to the Faculty and recommended for the degree by the Dean, who is, under our regulations, responsible for their having complied with the conditions prescribed for candidates for that degree. The candidate for the Doctor's degree will be presented by a representative of the Group Committee under whose jurisdiction his major work has been done. A sketch of the candidate's academic history will be read to the Faculty; and in case of the candidates for the Doctor's degree, there will be added to this sketch an outline of the thesis, giving in language intelligible to those who are not specialists in that line of work a clear idea of its scope and significance. Any member of the Faculty will then be at liberty to ask of the candidate or of the presenter any questions he may desire. After all the information relating to each candidate has been submitted, the Faculty will vote upon the question of recommending him for the degree in question.

To secure similar publicity outside the University, it has been provided: (1) That the statements read to the Faculty should be published *in extenso* in the *University Bulletin,* and given the widest possible circulation. (2) That the thesis shall be printed in full, and that 250 copies shall be delivered to the University. These will be distributed to institutions of learning, and to scholars familiar with the subjects of which the thesis treats, throughout the world. These provisions went into effect at the close of the academic year 1899-1900, students already in the Department being allowed to proceed to their degrees under the former rules, if they desired to do so. Of the fifteen candidates who then received the Ph. D. degree, six availed themselves of this permission, the remaining nine preferring to conform to the new provisions. All the eleven candidates for the A. M. degree chose the new conditions. The statements read to the Faculty as to the academic record of the candidates for the Ph. D. degree were printed in the Commencement number of the *Bulletin,* and an edition of 2,000 copies was distributed by mail. The printed theses also were distributed as provided in the rules.

If I have reported upon this matter somewhat more at length than is customary, it is because I feel that no one university should depart so widely from traditional usage without placing clearly before the academic world the reasons for such a step. Our aim has been the maintenance of the integrity and value of the higher degrees. Time only can show whether our action has been wise.

At the same time, certain other changes were made in our regulations, to which I need do no more than briefly refer. Candidates for the Ph. D. degree have hitherto been required to pass their major and minor examinations at about the same time. Hereafter they may pass the examination in any minor when it has been completed, and the examination in the major when all required resident and lecture work has been completed, including both major and minors.

Candidates for the Ph. D. degree have hitherto been required to complete the thesis before presenting themselves for the final examinations. Hereafter they may pass all examinations before completing the thesis, and may be allowed to complete the thesis *in absentia,* provided the conditions under which the work is to be done are satisfactory to the Committee in charge of the major.

We have hitherto required that candidates for the Ph. D. degree should have a reading knowledge of two European languages, one of them being a modern tongue. Hereafter French and German will be required unconditionally.

EQUIPMENT.

The Class of 1889 has contributed a sum of money, which will amount to about $2,000, for the purchase of books in Philosophy and kindred subjects. This is a most welcome gift. Our library of Philosophy is extremely defective. Our resources have not been sufficient to secure all the current books, much less to fill up the many gaps in the older literature. This the generosity of the Class of 1889 will enable us to do.

The Fellowship Funds provided by the George L. Harrison Foundation, which have amounted in this past year to $1,800, have been for the most part expended in the increase of our Library. That connected with the Fellowship in Botany has, however, been expended in the purchase of apparatus.

A generous friend of the University has provided the funds necessary for the equipment of the rooms in the Library building, formerly occupied by the Museum, as Seminary rooms. These rooms have been assigned to the various committees of instruction, and have been for nearly a year past in daily use. A system of departmental libraries has been established in them. These rooms are for the most part admirably adapted to their new uses, and will meet a need which has been long and painfully felt. Our only pressing needs in the way of material facilities are now additional laboratory accommodations in Botany, Zoölogy and Psychology. Details with reference to other additions to our equipment will be found in the reports of the Dean of the College and of the Librarian.

FELLOWSHIPS.

The temporary University Fellowship in English, established last year by Mr. J. F. Sinnott, has been continued by him for the year 1900-1901, and Dr. Joseph Vincent Crowne has been reappointed to it.

Of the five Senior Fellowships on the George Leib Harrison Foundation, three have been filled.

Dr. William Fairley, during his incumbency of the Senior Fellowship in European History, took up a "Constitutional History of the Later Roman Empire" as his theme. He hopes to cover the period from Diocletian to Justinian, and to make, probably, two volumes. During the year he succeeded in getting into practical readiness for the press one section of the first volume. That volume is to contain a preliminary chapter entitled a "Sketch of Three Centuries," which will give a swift review of the constitutional development from Julius to Diocletian. Then a very carefully worked out chapter, which will give in ample detail a cross-section of the "Constitution as Found by Diocletian." Upon this, two chapters of the changes under Diocletian and Constantine are to follow; and it is his intention to finish it within the present academic year. Dr. Fairley will be during the year 1900-1901 teacher of history in the Brooklyn High School.

Dr. E. S. Meade, Senior Fellow in Economics, has devoted himself to two pieces of original work, in both of which Mr. F. A. Cleveland, Harrison Fellow in Economics, collaborated with him. These were (1) Collection of Statistical Material for a History of Prices in the United States; (2) History of Railway Financiering in the United States. In addition to these he has published four articles, "The Relative Stability of Gold and Silver," *Annals of the American Academy of Social Science,* July, 1899; "Recent Production of Silver and its Probable Future," the same, November, 1899; "Stability of the Gold Standard," an address delivered at the meeting of the American Social Science Association, Washington, D. C., published in the Proceedings of the Association; "American Coal Supremacy," *Forum,* October, 1900.

Dr. Meade has also given one course of instruction in Economic Problems, two hours a week throughout the year, to Freshmen in Business Practice and Banking. During the year 1900-1901, Dr. Meade will be Instructor in Economics in this University.

Dr. Martin Schütze, Senior Fellow in Germanics, conducted two graduate courses, one on "The Storm and Stress Movement in German Literature," two hours for one term, and one on the "Contemporary German Drama from Hauptmann to the Present Day," two hours, two terms. Both of these courses of instruction represent original work which is ultimately to be issued in printed form. In connection with the first course, Dr. Schütze has collected an almost complete bibliography of the subject, and much of the material for the book. The material upon which the second course of lectures has been based is now being thrown into the form of a book, which will be completed about January, 1901. It will contain about 250 printed pages, and will comprise short biographies of all the dramatic poets and realists of the new movement, from Hauptmann's "Vor Sonnenaufgang" to Holz Schlaf's "Neue Geleise," with complete bibliographies and chronological lists of their works. Upon this material will be based a study of the main tendencies and social conditions from which this school arises, of the principal problems with which it deals, of the literary genesis of naturalism, the progress of the movement in other countries, and the peculiar technique of the naturalistic drama in Germany.

In addition to this work, Dr. Schütze translated Ludwig Fulda's play, "Friends of Youth," for the production of the German version of it under Dr. Conried's management for the benefit of the German Publication Fund. During the year 1900-1901, Dr. Schütze will again hold his Senior Fellowship.

Dr. J. V. Crowne, University Fellow in English, has continued his work upon the Early English Devotional Lyric. In particular he has been devoting himself to a fuller investigation of the Middle English Lyric with regard to facts treated more briefly in his thesis, with especial reference to their Latin, French and German parallels and sources. Dr. Crowne has also given four hours of instruction in the College in Public speaking and in English Composition to Freshmen and Sophomores.

During the past year Dr. Adeline F. Schively, Honorary Fellow in Botany, has been unable, owing to unexpected pressure of other work, to prosecute her research studies as actively as before. She is gradually extending her investigations, however, and hopes to incorporate these in a special paper. She gave several lectures before our Botanical Society, which were greatly appreciated. As her time will be fully occupied for the coming session, she has felt it to be her duty to resign the Honorary Fellowship held during the past three years.

Dr. S. C. Schmucker, Honorary Fellow in Biology, has conducted some interesting phyto-phenological studies, which must extend over several years to be of permanent value. He again delivered a course of three lectures to our students on "The Chemistry of Plants"—a subject he is eminently qualified to handle.

Of the nineteen Harrison Fellowships, eighteen have been filled, the appointee to the Fellowship in Semitics having been compelled to resign before the opening of the year by reason of ill health.

Of the eight Harrison Scholarships, seven were occupied, and of the thirty University Scholarships, twenty-nine were occupied.

Number of Students.

The registration for the past year shows a gain of thirteen upon that of the preceding year. The figures are as follows:

Total number of students registered, 177, of whom two withdrew without doing any work, leaving the net registration 175. Six of these were Senior, University and Honorary Fellows, having already taken their Ph. D. degrees. Thirty-one were auditors, who, holding no baccalaureate degree, cannot become candidates for a higher degree, leaving 138 students who were fitted to present themselves for candidacy for a higher degree. Of these 138 students, 66 were in the Department during the year before, while 72 were new students.

The subjoined table will give the registration figures for 1899-1900, the corresponding figures for 1898-99 and 1897-98 being annexed in brackets:

	New Students.			Old Students.			Total.		
	1899-1900.	98-9	97-8	99-00	98-9	97-8	99-00	98-9	97-8
Senior, University and Hon. Fellows	5,	[3,	1]	4,	[5,	6]	6,	[8,	7]
Regular students	72,	[68,	42]	66,	[62,	59]	138,	[130,	101]
Auditors	19,	[13,	21]	12,	[11,	32]	31,	[24,	53]
	93,	[84,	64]	82,	[78,	97]	175,	[162,	161

The following figures will show the number of women in the Department during the last six academic years:

	'94-5	'95-6	'96-7	'97-8	'98-9	'99-1900
Fellows and other Regular Students.	10	13	13	18	21	32
Special Students	24	17	14	18	11	12
	34	40	27	36	32	44

The percentage of women in the Department is shown in the following table:

	'94-5	'95-6	'96-7	'97-8	'98-9	'99-1900
Of the Fellows and other Regular Students	10	12	12	14	16	23
Of the Special Students	40	28	17½	30	46	33
Of the Total Number of Students	21	18	15	22½	20	25

Institutions Represented.

The following list will show the number of students in the Department during the year 1899-1900 from each of the following institutions:

Amherst 2
Beloit 1
Berlin 3
Bowdoin 1
Brown 1
Bryn Mawr 4

Buchtel	4
Bucknell	1
California	1
Catholic University of America	1
Chicago	9
Colorado College	1
Columbia	3
Cornell University	4
Delaware	1
Denison	1
De Pauw	1
Dickinson	4
Drury	1
Earlham	1
Franklin and Marshall	1
Georgia	1
Gymnasium of Güstrow	1
Gymnasium of Haarlem	1
Harvard	5
Haverford	4
Heidelberg	1
Hiram	1
Illinois	1
Indiana	4
Iowa	3
Jefferson	1
Johns Hopkins	2
Juniata	1
Kenyon	1
Lafayette	3
Lehigh	1
Leipzig	2
Michigan	3
Mount Holyoke	2
Muhlenberg	1
Koenigliches Gymnasium, Munich	1
Nebraska	1
Oberlin	2
Ohio Wesleyan	3
Pennsylvania	40
Princeton	1
Radcliffe	1
Real Schule, Hamburg	1
Rhode Island College	1
Rochester	1
Smith	1
South Dakota	1
University of St. Andrew	1
Stevens Institute	1
St. Joseph's	2
St. Lawrence	1
Swarthmore	4
Syracuse	1
Tennessee	1
Thiel	1
Toronto	1
Trinity	2
Tufts	1
Ursinus	5
Washington and Jefferson	1
Wellesley	1
Wells	1
Wesleyan	1
Williams	1
Wilson	2
Yale	3

Distribution of Courses.

The elections of students do not show any marked changes as compared with those of last year.

ELECTIONS OF REGULAR STUDENTS.

Min., abbreviation for Minor. Maj., abbreviation for Major.

	For the Doctor's Degree.												For the Master's Degree.					
	'94-95	'95-96	'96-97	'97-98	'98-99	'99-00	'94-95	'95-96	'96-97	'97-98	'98-99	'99-00	'94-95	'95-96	'96-97	'97-98	'98-99	'99-00
	Min.	Min.	Min.	Min.	Min.	Min.	Maj.	Maj.	Maj.	Maj.	Maj.	Maj.						
Comparative Philology and Sanskrit	2	3	3	4	5	5	1	...	...	1	1	1	...	...	...	...	2	2
Greek	6	7	7	6	6	5	2	1	2	4	5	5	...	2	1	...	4	5
Latin	2	2	2	4	4	6	1	2	3	3	7	5	2	2	...	1	5	5
Romanic	2	3	2	6	8	6	...	2	2	...	3	2	...	1	2	1	...	2
Germanic	1	2	1	1	4	5	...	1	1	...	5	8	...	...	2	2	...	2
English Philology	6	7	6	7	11	10	1	1	...	...	1	1	...	2	...	2	1	4
English Literature	6	7	10	11	5	3	3	5	4	7	13	14	1	2	3	4	3	5
Assyrian	2	3	1	1	1	2	3	4	4	8	1	...	...	...	...	...	...	...
Hebrew	7	7	6	6	2	...	7	2	3	2	2	4	...	...	1	...	...	...
Syriac	2	...	...	...	...	1	...	...	...	...	...	...	...	...	...	...	...	...
Arabic	3	2	2	1	2	...	...	...	...	...	1	...	...	...	...	...	1	...
Ethiopic	...	...	...	...	...	...	...	...	...	...	...	...	...	...	...	...	...	...
American Languages	...	...	...	...	...	...	...	...	...	...	...	...	...	...	...	...	...	...
Philosophy	31	29	26	15	14	8	16	14	9	8	4	5	2	5	3	1	1	3
Psychology	7	8	10	8	2	4	6	7	2	3	6	4	1	2	2	1	...	...
Ethics	9	8	9	5	4	4	1	3	4	1	...	...	2	2	1	...	...	...
Pedagogy	12	15	7	8	9	2	13	11	11	4	8	8	...	...	1	2	...	1
American History	3	6	11	6	9	8	7	8	4	6	9	7	2	2	1	...	2	2
American Constitutional History	...	2	3	3	1	1	...	1	2	1	1	1	1	1	...	...	...	...
European History	9	12	8	11	12	7	5	5	6	1	5	4	3	6	2	1	...	3
Political Science	7	9	7	4	4	5	4	6	4	2	1	2	...	1	...	...	1	...
Economics	27	24	14	9	6	7	3	4	6	5	7	5	1	2	...	1	2	2
Statistics	...	4	2	3	2	1	...	...	...	...	...	...	...	1	...	...	...	...
Mathematics	4	2	2	8	7	8	5	4	4	3	6	6	1	...	...	3	1	2
Astronomy	...	1	4	...	1	4	...	...	...	3	4	3	...	...	...	1	1	...
Physics	6	6	3	6	2	1	1	...	1	2	2	2	...	...	...	...	...	1
Chemistry	2	2	3	5	5	3	11	12	13	10	4	5	...	1	...	...	...	...
Geology	11	10	11	10	5	3	...	...	...	...	...	1	1	1	...	...	...	...
Mineralogy	8	7	7	7	5	3	...	1	1	1	...	1	...	...	...	...	1	...
Botany	3	2	1	4	5	5	3	1	2	1	3	3	...	2	...	...	...	...
Zoölogy	7	4	3	4	6	5	4	3	3	6	9	8	...	2	2	...	...	...
Italian and Spanish	...	...	2	...	1	2	...	...	...	...	...	...	...	...	...	...	...	1
Old Norse	...	...	1	...	2	1	...	...	...	...	...	...	...	...	...	...	...	...
Mediæval History	...	...	...	1	...	...	...	...	...	2	1	...	...	...	...	...	...	...
English History	...	...	...	1	2	1	...	...	...	1	...	...	...	...	...	...	...	...
Organic Chemistry	...	...	...	7	7	12	...	...	...	2	2	...	...	...	...	...	2	...
Inorganic Chemistry	...	...	...	4	2	1	...	...	...	5	6	10	...	...	...	...	2	...
Analytical Chemistry	...	...	...	3	3	1	...	...	...	...	...	...	...	...	...	...	1	...
Comparative Anatomy of the Vertebrates	...	...	...	1	1	...	...	...	...	...	...	...	...	...	...	...	...	...
Electro Chemistry	...	...	...	...	1	5	...	...	...	1	...	...	...	...	...	...	...	...
Sociology	...	...	...	...	4	5	...	...	...	1	3	7	...	...	...	...	...	2
Minors not yet elected	...	...	10	9	70	92	...	...	...	...	...	...	...	...	...	...	...	...
	185	194	182	188	240	242	97	98	91	94	120	122	17	37	21	20	30	42

ELECTIONS OF SPECIAL STUDENTS.

	1894-95	1895-96	1896-97	1897-98	1898-99	1899-00
Comparative Philology and Sanskrit	1	—	—	—	—	—
Greek	1	1	1	2	1	—
Latin	1	3	2	4	3	2
Romanic	1	1	2	2	2	3
Germanic	—	3	2	2	1	—
English Philology	2	—	2	—	1	2
English Literature	8	9	8	6	4	4
Assyrian	2	5	1	1	—	—
Hebrew	4	7	6	2	—	—
Syriac	—	—	—	—	—	—
Arabic	2	3	1	—	—	—
Ethiopic	—	—	—	—	—	—
American Languages	1	—	—	—	—	—
Philosophy	16	11	14	6	4	5
Psychology	6	6	17	9	2	4
Ethics	2	5	6	5	—	3
Pedagogy	8	9	23	15	10	8
American History	7	9	6	5	2	—
American Constitutional History	5	5	6	2	2	—
European History	7	7	6	—	—	—
Political Science	—	1	3	1	2	—
Economics	2	4	3	2	1	1
Statistics	—	—	1	—	—	—
Mathematics	5	3	4	1	2	2
Astronomy	—	1	1	—	—	—
Physics	—	1	3	2	—	1
Chemistry	1	1	7	2	1	3
Geology	1	3	3	1	—	—
Mineralogy	1	1	2	—	1	1
Botany	2	2	3	1	—	1
Zoölogy	2	3	1	3	—	—
Biology	—	—	2	—	—	—
Sociology	—	—	1	—	1	3
	88	104	137	74	40	43

DEGREES.

At the Commencement held on June 13, 1900, degrees were granted as follows: In connection with each student's name will be found his subjects, and his address for the year 1900-1901. In case of candidates for the Ph. D. degree, the major comes first, and the title of the thesis is also given.

Master of Arts.

Mary Maxwell Blaine: Mathematics, Physics.
Teacher of Mathematics, High School, Springfield, Mo.

Frank Goess Bossert: Philosophy, European History, Sociology.
Student, Princeton Theological Seminary.

Christian Carl Carstens: Sociology, Economics, American History.
Assistant Secretary of the Charity Organization Society, Philadelphia, and continues graduate work in the University of Pennsylvania.

Daniel Ernest Martell: Romanic Languages, German.
Continues graduate work, University of Pennsylvania.

Alice Madeleine McKelden: Mathematics, Greek, Classical Philology.
Teacher of Latin and Mathematics, Stonewall Jackson Institute, Abington, Va.

John Alvin Orr: American History, English Literature, European History.
Student, Theological Seminary of the United Presbyterian Church, Allegheny, Pa.

Mary Jane Ross: Zoölogy, Chemistry.
Assistant in Zoölogy and Embryology, Cornell Univ.

Edward Mitchell Spencer: English Literature, English Philology, European History.
Teacher, Hudson River Institute, Claverack, N. Y.

Henry Wilson Stahlnecker: English Literature, English Philology, European History.
Student, Law Department, University of Pennsylvania.

Edward Charles Wesselhoeft: Germanics, Romanic Languages.
Instructor in German, University of Pennsylvania.

Milton Bigler Wise: European History, Philosophy, Pedagogy.

Continues graduate work in the University of Pennsylvania.

DOCTOR OF PHILOSOPHY.

William Harvey Allen, A. B. (Chicago, 1897): Political Science, Economics, American History.

Thesis: *Rural Sanitary Administration in Pennsylvania.*

Instructor in Public Law, University of Pennsylvania.

Barclay White Bradley, A. B. (Pennsylvania, 1897): Greek, Latin, Comparative Philology.

Thesis: *The Use of the Chorus in the Dialogue Portions of Sophocles.*

Instructor in Classics, Temple College, Philadelphia.

Frederic Albert Cleveland, Ph. B. (De Pauw, 1890): Economics, Political Science, Finance.

Thesis: Statistical materials for four chapters of a work to be entitled, *History of Prices Since 1860.*

Instructor in Finance, University of Pennsylvania.

James Edward Hagerty, A. B. (Indiana, 1892): Sociology, Economics, Political Science.

Thesis: *Recent Changes in the Marketing of Products in Their Effects upon Social Welfare.*

Senior Fellow in Sociology, University of Pennsylvania.

Orlando Faulkland Lewis, A. B., A. M. (Tufts, 1895, 1897): Germanic Literature, Germanic Philology, English Philology.

Thesis: *History of German Drama in Philadelphia, Part I, Statistical.*

Assistant Professor of German, University of Maine.

Morton Githens Lloyd, B. S. in E. E. (Pennsylvania, 1896): Physics, Mathematics, Chemistry.

Thesis: *The Transversal Thermo-magnetic Effect in Bismuth.*

Instructor in Physics, University of Pennsylvania.

John Damien Maguire, A. B. (La Salle, 1886): Latin, Greek, Comparative Philology.

Thesis: *De Verborum in Livianis Orationibus Collocatione.*

Assistant Professor in the Department of Latin, Catholic University of America, Washington, D. C.

Anna Jane McKeag, A. B. (Wilson College, 1895): Psychology, Philosophy, Philosophy.

Thesis: *The Sensation of Pain: an Experimental and Critical Analysis.*

Professor of Philosophy, Psychology and Ethics, Wilson College, Chambersburg, Pa.

Albert Edward McKinley, Ph. B. (Chicago, 1896): American Constitutional History, American History, Political Science.

Thesis: *Representation and the Suffrage in New Netherlands and New York, 1613 to 1691.*

Professor of History and Political Economy, Temple College, Philadelphia; Honorary Fellow in American Constitutional History, Pennsylvania.

George Ward Rockwell, B. S. (Buchtel, 1898): Inorganic Chemistry, Organic Chemistry, Electro-Chemistry.

Thesis: *An Electrolytic Study of Pyroracemic Acid.*

Chemist to the Bonneville Cement Company, Siegfried, Pennsylvania.

Charles Lawrence Sargent, B. S. in Agriculture. (R. I. College of Agriculture and Mechanic Arts, 1894): Electro-Chemistry, Organic Chemistry, Inorganic Chemistry.

Thesis: *Alloys of Tungsten and of Molybdenum obtained in the Electric Furnace.*

Chemist to the Murphy Varnish Company, Newark, New Jersey.

Charles Hugh Shaw, B. S. (Ohio Wesleyan, 1897): Botany, Botany, Zoölogy.

Thesis: *A Comparative Study of the Flowers of Polygala polygama and P. pauciflora, with a Discussion of Cleistogamy.*

Professor of Biology, Temple College, Philadelphia, and Honorary Fellow in Botany, University of Pennsylvania.

Lewis S. Shimmel (Graduate of Millersville State Normal School, 1877). Admitted to candidacy by special action of the Executive Committee, 1896: American History, American Constitutional History, Pedagogy.

Thesis: *Border Warfare in Pennsylvania during the Revolution.*

Teacher of United States History and Civil Government, Harrisburg (Pa.) High School.

Claude Halstead Van Tyne, A. B. (Michigan, 1896): American History, European History, Political Science.

Thesis: *The Elimination of the Loyalists by Legal Enactments,* chapters 8, 9 and 10 of a larger work to be entitled, *Treatment of the Loyalists during the American Revolution.*

Senior Fellow in American History, University of Pennsylvania.

Albert Duncan Yocum, Ph. B. (Dickinson, 1889): Pedagogy, Psychology, European History.

Thesis: *An Inquiry into the Fundamental Processes of Addition and Subtraction.*

Superintendent of Schools for the City of Chester, Pa.

NEEDS.

We at present need, above all things else, additional instruction in subjects which are now either inadequately represented, or are not represented at all in our graduate work. Some members of our teaching staff are also overburdened with teaching and administrative work, and need more time for the preparation and publication of original work. Such work it is, already done by the instructors of this Department, which is drawing our brightest students; and upon such work alone the prestige of our graduate instruction will in the last analysis be found to rest. Yet I know, Sir, that there is no need of my calling this aspect of our work to your attention. Your deep interest in it would have been sufficiently attested by the conditions laid down in the establishment of the George Leib Harrison Foundation, even if it had not been evinced in so many other ways. I realize that our present circumstances make it necessary for us to fall short in many particulars of the ideal which we have set for ourselves, and towards the realization of which, through the acquisition of an adequate endowment, the efforts of the administration are unceasingly directed.

Respectfully submitted,

Wm. Romaine Newbold,

Dean.

APPENDIX IV.

REPORT OF THE DEAN OF THE DEPARTMENT OF LAW.

To the Provost of the University;

SIR:—I beg to submit herewith my annual report for the year ending August 31, 1900:

STUDENT STATISTICS.

The total number of students in attendance during the past year was 313: 92 in the third-year class, 79 in the second-year class, 120 in the first-year class, 9 special and 13 partial students. These figures show an increase in the entering class as compared with the fall of 1898 of 29 students. There was a decrease of 7 in the total registration. This decrease, in spite of the increase in the size of the entering class, was due to the fact that the graduating class in June of 1899 contained 114 men, this class being the last to enter under the old entrance requirements.

We have more than recovered from the effect of the changes in our entrance requirements. These changes were announced in the winter of 1898. In 1895 and 1896, the registration in the Department was 273, the first-year class numbering 111.

Two hundred and seventy-six students registering last year came from Pennsylvania, 185 of this number being from Philadelphia; 35 came from States other than Pennsylvania, and 2 from foreign countries. Ten States were represented. Ten years ago there were 94 students from Pennsylvania, 85 of which were from Philadelphia, and only 8 from States other than Pennsylvania.

The average age of the students in the Department was

twenty-two years and four-tenths, the average age of the entering class being twenty years and seven-tenths.

Thirty-eight per centum of all the students and 41 per centum of the first-year class were college students; 20 per centum of all these students and 16 per centum of the first-year class had attended college for one or more years, but had not graduated from college; 23 per centum of all the students and 31 per centum of the first-year class were high school graduates; 18 per centum of all the students and 11 per centum of the first-year class came from private preparatory schools, or had been prepared for the entrance examinations by tutors. These figures show that the proportion of college graduates and high school graduates has increased, while the proportion of what may be called partial college men, and of the non-college and non-high school class, has decreased. The college graduates represented forty-one institutions, the largest number, twenty-eight, coming from the College of this University. The representation of Princeton came next, fourteen of our students having graduated at the College of that University. There were ten students from Yale, and nine from Harvard and Haverford, respectively. It is interesting to note the increase of college graduates registered in the Department in the last ten years. In 1889-1900 we had thirty-two college men, while during the past year we had one hundred and twenty-six, or an increase of nearly three hundred per centum. During the same period the number of colleges represented increased from thirteen to forty-one, as above stated.

Results of the Examinations in Course.

The average of all the marks received by the students in the Department in the June examinations was 74 9-10 on a basis of 100. Out of 1,775 examinations, 345 honors or marks over 90 were given, and 196 conditions or marks under 65. The average of the marks received by those representing the four Universities of Pennsylvania, Harvard,

Yale and Princeton, was 81, the average of the marks received by all college graduates being 79 9-10. The average of the marks received by the partial college men was 74 9-10, of the high school men 76 7-10, and of all others 74 6-10. Of the college men, 28 per centum received honors and 7 per centum conditions; of partial college men, 16 per centum received honors and 4 per centum conditions; of the high school men, 16 per centum received honors and 13 per centum conditions; and all other men, 11 per centum received honors and 14 per centum conditions. These figures show that, as in former years, the college graduate as a class has proved to be the best student. Of the first ten men in the class graduating last June, seven were college graduates. On the other hand the examination marks of the men who have only spent a year or more at college, but not graduated—that is partial college men—show that they contain among them a fair proportion of good men, but that taken as a class they are less capable than high school men. The high school men show themselves more capable than those from private preparatory schools or from tutors.

Scholarships.

Fourteen men registered in the Department held scholarships, the total assistance given amounting to $2,240. Of these, six were free Faculty scholarships, four were free City scholarships and four were City scholarships paying to the University $100 each.

Faculty Changes.

Mr. James Parsons, Emeritus Professor of Law, died on the twenty-second of March last. Professor Parsons was elected a Professor of Law in 1870, and discharged the duties of his professorship until 1897. At the time of his death he held the honorary position of Emeritus Professor. At a meeting held a few days after his decease, the Faculty of Law passed a resolution, in which they accurately stated that they had lost a true and trusted friend, and that the

University had lost a faithful servant. Professor Parsons was a scholar widely known for his erudition and his writings.

It is also my painful duty to mention the death of our Instructor, Roy Wilson White. Mr. White graduated from the Department in 1898, dividing the honor of standing first in his class with Mr. Roberts. Together with Mr. Roberts, he was elected a Fellow in the Department, and at the end of the first year was sent by this University to the University of Paris for the purpose of continuing his studies in the Civil Law. He returned to Philadelphia in February, and devoted the majority of his time to preparing for the work of the Graduate School, in which he was to have conducted the Course on Comparative Jurisprudence. He was killed while returning to his home from the Law School building on the evening of the nineteenth of May. He was a man of rare ability, with a marked faculty for teaching. His capacity for hard work, his perseverance and the singular purity of his character, had endeared him to all the members of the teaching force; and we feel that it will take a long time even to begin to supply his place.

Mr. Arthur Dickson, the Senior Fellow, was obliged to resign in March, having removed to Lancaster, Pa. The Faculty accepted his resignation with regret, as his work had been most efficient. Mr. Thomas R. White and Mr. Arthur Weil, who were the Junior Fellows, performed the duties assigned to them in a manner most acceptable to the Faculty. Mr. Charles L. McKeehan, of the Class of 1900, was recommended by the Faculty and appointed by the Trustees as Fellow for one year, the term of service to begin on the first of September; while Mr. Thomas R. White has been elected a Lecturer in the Department.

Course of Instruction.

The new course in the first year last year was the course in Blackstone, or Elementary Law, two hours each week throughout the year. The course in Domestic Relations was dropped, as it was found on investigation that all the subjects taught in the course, except questions relating to guardian and ward, were taught in the other courses; and questions relating to guardian and ward could be properly included in the course in Blackstone. In the second year, a course in the Practice of the Orphans' Court of Pennsylvania was added, this course being taken only by men expecting to practice in this State. Those expecting to practice in other jurisdictions had the choice of taking an elective subject, or of taking an equivalent amount of the practice of the jurisdiction of the State in which they expect to locate. This change was due to the belief that there is no reason why Practice should not be taught at least throughout the last two years of the course, in conjunction with questions of Substantive Law.

The principal change in the third year was the addition of the new course on the Law of Association, given by Professor Pepper. This course is given as a substitute for the separate courses on Partnership and Corporations; and the change was in accordance with the general policy adopted by the Faculty during the last few years, of bringing under a single name, and in a continuous course, subjects of the law nearly related to each other. Another new elective in the third year was a course on New Jersey Practice, given by Mr. Norman Grey, of the Camden Bar. This elective is designed for the increasing numbers of our students who expect to practice in the State of New Jersey. It has been so successful that we have determined to make it during the coming year a two-hour course throughout the year, and to have it include the Statutes as well as the Practice of the State.

POST-GRADUATE COURSE.

During the spring an agreement was made between the Department of Philosophy and the Department of Law, which agreement was ratified by the Board of Trustees, looking towards the establishment of a course leading to the degree of Master of Laws. The subjects to be comprised in the course were Comparative Law, History of the Common Law, the Study of the Problems of the Common Law, and work in Government and Public Law now given in the Department of Philosophy. The course was to be open to the graduates of any three-year law school in good standing; the degree of Master of Laws to be conferred on the successful completion of a course amounting to at least ten hours each week for one year, the course on the problems of the Common Law involving not only an examination, but original work in the preparation of an essay. The sudden death of Mr. White, occurring as it did late in the spring, made it impossible for us to offer this year a course in Comparative Law. In view of this fact, and in view of the fact that Mr. White was not only to give this course, but to devote a great deal of his time to the necessary supervision of the graduate work in the Department, we were obliged, though reluctantly, to postpone the commencement of the course until the fall of next year. It is my earnest hope that we will be able, during the coming year, to make arrangements to enable us to establish in a proper manner a graduate course for the scholastic year 1901-1902. There is unquestionably a sincere desire for such a course among a large number of our alumni, and indeed among the younger members of the Bar in Philadelphia and elsewhere.

The Biddle Library.

The accession number of the Biddle Library on September 30, 1899, was 21,485. At the present time, September 30, 1900, it is 22,983, or an increase during the year of 1,498. This increase is divided as follows: Text-books, 425; Reports, 424; Statutes, 72; Digests, 47; Periodicals, 185; Bar Association Reports, 215; Ordinances, 108; Public Documents, University Documents and General Literature, etc., 14.

The removel of the Library from the old location at Sixth and Chestnut streets to the new building in West

Philadelphia, in itself would have caused some change in the details of library work. But the arrangement of the new Library is so radically different from that in the old building as to cause a complete change, not only in the details, but in the whole management. The separation of the Library into two reading rooms, with two delivery desks in the place of one, necessitates a greater number of attendants, even if there were no great increase in the work such as has taken place.

After moving to the new building a considerable increase in attendance was noted, and it is believed that during the coming year this increase will be even more marked.

We have already assigned twelve of the desks in the McMurtrie Graduate Reading Room to members of the Bar who desire to use the Biddle Library for original legal research.

The Dedication of the Law School Building.

The ceremonies connected with the dedication of our new building took place on the twenty-first and twenty-second of February. They were of a character befitting the importance of the event to the legal profession in the country and State. The building is not only the largest devoted exclusively to the study of the law in the United States, but the completeness of its arrangements, its dignity and the beauty of its interior decoration, make it one of the most complete educational buildings in the world. There has perhaps never been in this country such a gathering of eminent judges and lawyers and college dignitaries. The Supreme Court of the United States was represented by the Hon. John Marshall Harlan, the Senior Associate; the Universities of Oxford and Cambridge by Sir Charles Arthur Roe and the Hon. G. B. Finch, respectively; thirty universities and colleges were represented by their presidents; and there were also special representatives from all the leading law schools of the country. More than sixty State and County judges of Pennsylvania, besides a large number of

Federal judges from other States, and representatives of the various Bar Associations, were present as the special guests of the University; while the occasion gathered together many lawyers from all parts of the country. Besides the Reception Committee of the Pennsylvania Bar Association, the University was assisted in the entertainment of its guests by special committees of the Law Association of Philadelphia, the Lawyers' Club of Philadelphia, and a large committee composed of all the members of the Bench and the more prominent members of the local Bar.

At 1.30 on February 21, the University of Pennsylvania gave a reception at the new building. Lunch was served in McMurtrie Hall, and the rest of the building was thrown open to the inspection of the five thousand persons attending. You, together with the Trustees and Faculty of the Department, received in Sharswood Hall, one of the two large student reading rooms in the Law School building. The dedicatory exercises were held on the same afternoon at three o'clock in the other large reading room, McKean Hall.

The Right Reverend Ozi W. Whitaker, Bishop of the Diocese of Pennsylvania, offered the opening prayer. You, Sir, presided, making a short address of welcome. Mr. Samuel Dickson, the chairman of the Committee of the Board of Trustees on Law and Legal Relations, made an address, in which he presented the building, on behalf of the Trustees, to the Faculty of the Department of Law. I, as Dean of the Law Faculty, accepted the building on behalf of the Faculty. The closing address, on "The Vocation of the Law Professor," was made by Mr. James Barr Ames, the Dean of the Harvard Law School.

After the dedicatory ceremonies, the special guests of the University were entertained at private dinners given by the members of the University Club, the members of the Board of Trustees, the Faculty of the Department and the Bar.

At 8.30 p. m. there was a meeting at the American Academy of Music, at which an address was made by the

Hon. John Marshall Harlan, on "The Constitution of the United States"; also one by Sir Charles Arthur Roe on "The Constitutional Relations of England and Her Dependencies."

At eleven o'clock a reception was given by the University Club. The following day, Thursday, February 22, being University Day, the special guests of the University in attendance at the Law School opening attended the exercises in the Academy of Music, which took place at eleven o'clock in the morning.

The afternoon of Thursday was devoted to the inspection of the University grounds and buildings, and receptions given by the Deans of the various Departments in their respective offices, after which the Pennsylvania Debating Union gave a reception in the Law building. The guests then proceeded to Price Hall to listen to the address of Professor Hampton L. Carson, dedicating this hall, which is on the first floor of the Law School building, and is used by the Debating Union. Mr. Carson was followed by the Hon. G. B. Finch, who delivered an address on the "Development of the Common Law."

The two days' ceremony was closed with a commemorative dinner, given in Horticultural Hall under the auspices of the Law Association of Philadelphia, the Lawyers' Club of Philadelphia and the Bar Association of Pennsylvania. Nearly six hundred and fifty persons sat down to dinner. The walls of the banqueting hall, and the handsome staircase leading to it, were adorned with the portraits of eminent lawyers and judges. Mr. Samuel Dickson, Chancellor of the Law Association, and chairman of the committee in charge of the dinner, presided. Hon. William U. Hensel, of Lancaster, acted as toastmaster. Mr. Dickson, the Hon. George Gray, Judge of the United States Circuit Court; Sir Charles Arthur Roe, of the University of Oxford; Dr. Gerald B. Finch, of the University of Cambridge; Professor George Wharton Pepper, of the University of Pennsylvania; President Francis L. Patton, of Princeton University; Mr. James E. Parsons, of New York, and Mr. Richard C. Dale, of Philadelphia, made addresses appropriate to the occasion.

Gifts.

Several oil paintings have already been placed on the walls of the new building. The portrait of Mr. Richard C. McMurtrie, painted by William Chase, and presented by Mr. W. W. Frazier, hangs in the McMurtrie Reading Room. The portrait of Thomas McKean, painted by Vonnoh, presented to the Department in 1899, and temporarily deposited in Independence Hall, has been transferred to McKean Hall, the north student reading room. Professor Hampton L. Carson has presented a portrait of Mr. Justice Wilson, the first Professor of Law in the University. In connection with this gift, Mr. Thomas H. Montgomery presented a number of valuable original documents connected with Judge Wilson, which have been suitably framed, and now hang on the walls of the main hall of the Law School building. Among these documents is the letter of President Washington forwarding the commission of Wilson as Associate Justice of the Supreme Court of the United States, together with the original commission; his commission signed by Louis XVI. as Avocat-General; and his appointment as Professor of Law in the University of Pennsylvania.

The Class of 1900 established a custom, which it is hoped other classes will perpetuate. They presented, on their graduation, a handsome clock for the students' conversation room.

Law Clubs.

In 1881 the first students' law club, known as the Sharswood Club, was established. Since that time other clubs have been organized; the Hare Club in 1890, the Miller Club in 1894, the Kent Club in 1896. During the last year three new clubs were established—the Phi Delta Phi, the Wilson, and McKean. All these organizations are voluntary. The cases argued are prepared by the professor.

Each professor sits at least once during the winter for each club. As a consequence, every night in the week during the fall and winter there is always a moot court case being argued before a professor. The arguments may be attended by any member of the student body. I believe that the benefit derived by the students from these clubs is very great, and that this benefit is not solely due to the practice in the conduct of cases which they afford their members; for these clubs are also social organizations, and the life of the student body in a large measure centres around them.

RECEIPTS AND EXPENSES.

Receipts from tuition fees		$45,495 43
Receipts from miscellaneous sources		706 05
		$46,201 48
Paid to Treasurer of University	$40,568 00	
Biddle Library Fund	5,547 00	
	$46,115 00	
Balance on deposit	86 48	
		$46,201 48

The account of the Biddle Library Fund is as follows:

Cash on hand from 1899	$93 46	
Received as twelve per cent of receipts	5,532 00	
Donation	15 00	
		$5,640 46
Paid for the purchase of books		5,363 75
Balance on hand September 1, 1900		$276 71

The following is a detailed account of the expenses of the Department exclusive of the Biddle Library:

Educational salaries	$27,899 36	
Administrative salaries	2,478 72	
Current expenses, bookbinding, advertising, petty cash, wages, supplies, heat, light, repairs, etc.	9,674 62	
Total expenses		$40,052 70

Respectfully submitted,

WILLIAM DRAPER LEWIS,

Dean.

APPENDIX V.

REPORT OF THE DEAN OF THE DEPARTMENT OF MEDICINE.

To the Provost of the University;

SIR:—I have the honor of submitting to you the following report of the Department of Medicine for the academic year 1899-1900:

The students in attendance during the session, arranged in classes, numbered as follows:

Fourth-year Class	189
Third-year Class	164
Second-year Class	181
First-year Class	141
Special	4
	679

During this session there was a decrease of 156 in the total number of students in attendance upon instruction. In the session of 1898-99 the total number of students in attendance was 835; whereas, during this session, the total number of students in attendance was 679. The decrease in the number of students in attendance which has occurred annually during the past few years may be attributed to several causes:

First.—To the progressive annual increase in the severity of the requirements for admission to the first-year class.

Second.—To the withdrawal of students, because of failure to pass examinations in course, or for other causes.

Third.—To the increase of the tuition fee from $150 to $200.

Under the minimum entrance requirements in force for many years, viz.; (1) to write an essay of about three hundred words as a test of orthography and grammar; and (2) to pass an examination in Elementary Physics (Part I of Fowne's Chemistry): the first-year classes in the following-named sessions, immediately preceding the session in which increased entrance requirements were demanded, numbered:

Session 1894-95	242
Session 1895-96	333

The class that entered at the beginning of the session 1896-97 was admitted under increased requirements. These requirements were further increased with the class which entered with the session beginning 1898-99; and again increased with the class which entered with the session beginning 1899-1900, when the maximum increase of requirements, *i. e.*, the requirements for admission to the Freshman Class of the College of this University, were reached. Under these latter provisions all of the applications for admission to the first-year class, with the exception of the applications of individuals who had passed the examinations for admission to the Freshman Class of a college recognized by this University; or who had passed the examinations at the end of the Freshman or Sophomore or Junior year of a recognized college; or were graduated from a recognized college, were submitted for decision to the Committee on Admissions of the College Faculty: and the decision of this committee was final in each case. Those candidates whose credentials were considered unsatisfactory by the committee were obliged to undergo examination, in which examination the same questions were used that were used in the examination for admission to the Freshman Class of the College of the University; and the answers were submitted to the College Committee on Schools and Entrance Requirements. The decision of this committee was final in each case. The following table shows the size of the classes which entered at the beginning of the sessions 1894-95 to 1899-1900, inclusive:

		Members first-year class.
Old requirements:	1894-95....	242
	1895-96....	333
First increase of entrance requirements:	1896-97....	283
Second increase of entrance requirements:	1897-98....	205
Third increase of entrance requirements:	1898-99....	227
Maximum requirements:	1899-00....	141

The following table shows the number of candidates who underwent examination for admission to the first-year class in the academic years stated; and also shows the number who passed the examination, those who failed, and those who withdrew before completing the examinations:

Sessions.	Examined.	Passed.	Failed.	Withdrew before being examined in all studies.
1896-97	92	81	6	5
1897-98	37	21	7	9
1898-99	28	16	9	3
1899-00	22	15	7	—

The above table does not show the number of candidates who made application for admission on certificates from preparatory, high and normal schools, etc. The certificates of preparatory schools are in no way recognized for admission to the first-year class. Diplomas of graduation from high and normal schools are recognized, either as complete or partial certificates for admission. When recognized as a partial certificate, the candidate is obliged to undergo examination in the subjects not covered by his diploma.

There has been a gratifying increase in the percentage of candidates admitted to the first-year class who possessed collegiate degrees. While 22 of the 227 of the first-year class of the session of 1898-99, or 9.6 per cent, possessed collegiate degrees, the number of members of the first-year class of the session 1899-1900 who possessed collegiate degrees was 28 in a class of 141, or a percentage of 19.9, an increse of slightly over 10 per cent. If to the number of members of the first-year class possessing collegiate degrees were added the number (26) of college graduates who entered the second-year class on biological and scientific requirements, the total would be 54, or 38.3 per cent.

The status of general education of the members of the first-year class who entered the Department this session, compared with those who entered last session, may be shown by the following table:

	Session 1899–1900.	Session 1898–99.
Candidates who passed entrance examinations.	4	23
Candidates who took part of the examinations (remainder covered by diplomas from High or Normal Schools)	4	—
College graduates	28	22
Graduates of approved High Schools and Academies	40	93
Graduates of approved Normal Schools	3	12
Candidates who had been members of the Senior Class in a College of standing	1	3
Candidates who had been members of the Junior Class in a College of standing	3	6
Candidates who had been members of the Sophomore Class in a College of standing	11	24
Candidates who had been members of the Freshman Class in a College of standing..	13	19
Dropped from the preceding First-year Class and repeating the studies of that year....	34	25
	141	227

It may be observed by the table that 34 of the 141 members of the class were repeating the studies of the year, as against 25 out of a class of 227 in the preceding year. This may be attributed to an increase in the severity of the examinations in course which, to my mind, has occurred since the higher entrance requirements have been demanded.

The average age of the members of the first-year class was 21.1. The following tabulated statement shows the average ages of the members of the class who entered under varying conditions of preparatory education:

8

	Session 1899–1900.	Session 1898–99.
The average age of the members of the First-year Class was	21.1	20.7
Average age of College graduates	23.0	22.6
Average age of members who entered from the Senior Class of a College of standing	22.0	22.0
Average age of members who entered from the Junior Class of a College of standing	22.0	21.2
Average age of members who entered from the Sophomore Class of a College of standing	20.4	20.6
Average age of members who entered from the Freshman Class of a College of standing	20.8	19.9
Average age of graduates of High Schools, Academies, etc.	20.4	19.0
Average age of graduates of Normal Schools	24.3	....
Average age of those entering by examination	19.4	19.4

The following tabulated statement shows the status of the membership of the class at the end of the session after the examination periods:

Number entering session 1899-1900			141
Number in Second-year Class 1900-01 (succeeding session)		125	
Less number admitted to advanced standing	15		
Less number repeating second year from a former session	6		
		21	
Number advanced from first to second year		104	
Number repeating first year		4	
Number withdrawn		33	
			141

The figures in the following table show the number of members of the fourth-year class who selected the respective electives:

Nervous Diseases and Electro-Therapeutics	5
Skin Diseases	29
Advanced Hygiene (including Bacteriology)	0
Laryngology and Rhinology	38
Pharmacodynamics	1
Advanced Pathology	1
Orthopedic Surgery	11

Ophthalmology	14
Otology	4
Genito-Urinary Surgery	72
Advanced Anatomy	0
Advanced Physiology	0
Advanced Medical Chemistry (including Toxicology)	0
Pediatrics	67
Experimental Psychology	0

The rearranged curriculum, with special reference to the third and fourth years of the course, which went into effect with the beginning of this session, provided much more practical clinical work than was heretofore given in the course; and was in general an admirable improvement upon the older methods. In the rearranged curriculum, in addition to the changes noted in the last report of this Department, the lectures on the History of Medicine were transferred from the first to the fourth year of the course; the lectures on Physical Diagnosis and on General Symptomatology in the first year were abolished; instead of didactic lectures twice a week on Applied Anatomy throughout the entire session, demonstrations in that subject were given three times a week; and the third-year class was divided into three sections, corresponding with the number of demonstrations. Likewise, in the teaching of Pharmacodynamics, the class was divided into three sections, corresponding to three weekly demonstrations. In the teaching of Chemistry of the first year, and Medical Chemistry of the second year of the course, each of the classes was divided into three sections, each section having three consecutive hours a week in the Laboratory during the entire session.

Extensive improvements have been made in the Dissecting-Room by the installation of an adequate number of marble-top washstands supplied with hot and cold water; and the painting of the walls and ceiling of the room, so that there should be the least absorption of light.

The general clinic in Gynecology was abolished, and in its stead was substituted a lantern demonstration by the professor of Gynecology.

At the request of the Faculty of Medicine, the Board of

Trustees consented to the discontinuance of the practice of admitting college graduates on biological and scientific qualifications to the second-year class. This action will go into effect with the session beginning 1901-02. As the result of conferences between committees from the Faculty of Medicine, and from the Faculty of the College, a composite year has been arranged: so that members of the Senior Class of the College may matriculate in the Department of Medicine at the beginning of their Senior year in the College, and take the studies of the first year of the Department of Medicine in that Department, in conjunction with certain studies of the Senior year of the College. On passing the required examinations in the College, and also in the Department of Medicine, they will be recommended for the Bachelor's degree, and may subsequently proceed with the studies of the second year of the course in the Department of Medicine. This plan will enable the student to obtain the Bachelor's degree and the degree in Medicine within a period of seven years.

The members of the Faculty are convinced that the accommodations for teaching many of the studies of the first and second years are inadequate, and that it is exceedingly desirable that the proposed building to accommodate the laboratories of Pathology, Physiology and Pharmacodynamics should be completed as soon as practicable.

Respectfully submitted,

JOHN MARSHALL,
Dean.

APPENDIX VI.

REPORT OF THE UNIVERSITY HOSPITAL.

To the Provost of the University;

SIR:—I have the honor of transmitting the report of the Board of Managers of the University Hospital as submitted at the last annual meeting. It is as follows:

In this, the twenty-fifth annual report of the Hospital of the University of Pennsylvania, the Managers point with pleasure to the increased usefulness of the institution, and to the many additions and betterments which have been made in 1899. The past year has witnessed many changes in the administration of the Hospital. The former Superintendent and Assistant Superintendent having resigned, Miss Jessie W. Brown was elected Superintendent, and Miss Jean Macpherson, Assistant Superintendent. A new office has been created in the Residential Staff, viz., that of Chief Resident Physician, his duties being to take charge of and to be responsible for the Receiving Ward; and in the absence of the Physicians-in-Chief, to take charge of the medical and surgical administration of the Hospital. Dr. M. H. Biggs has been appointed to fill this position. Through the liberality of J. Dundas Lippincott, Esq., a new operating room, with bathroom, etherizing and recovery rooms, has been built for the use of private patients. Owing to the fine northern exposure, and to bringing the skylight down close to the floor, a light has been obtained far superior to that obtained by most northern exposures, while the steam heating is so placed as to prevent any chill to the patients. Of course the water sterilizers, the sterilizers for dressings, and the sterilizers for instruments are conveniently placed.

The whole is given as a memorial to the late James Dundas, and is entitled "The James Dundas Operating Room for Private Patients." The same generous donor has given a complete outfit of instruments, and an outfit of linen. The new Nurses' Home has been completed, and the old building remodeled. There are now comfortable accommodations for seventy-five nurses, with pleasant bedrooms, sitting-rooms and ample bathing facilities. A new filter has been placed in the cellar of the Home, of size sufficient to clarify every drop of water used for drinking, cooking or bathing. A large gas stove and gas broiler have also been placed in the building, furnishing a most agreeable method of cooking in warm weather, as the heat lasts only as long as there is a necessity for it. The new asphalt walk between the Hospital and the Nurses' Home makes the transit in wet weather drier and more salubrious. The system of nursing has also been changed, and at present we have a *graduate head nurse* in each ward, thus providing for continuously efficient nursing. Each subordinate nurse has an opportunity to learn her business under the most careful and competent supervision, while the chance of any injury by carelessness or ignorance is reduced to a minimum.

The changing and improving of the uniforms of the nurses, the uniforming of the ward maids and of the door boys, have been real improvements in the discipline of the Hospital. A typewriter in the office of the Hospital has also added to the efficiency of this department. The quarters of the Resident Physicians have been made more comfortable by a new tiled bathroom and a white porcelain tub. Similar changes have been made in the bathrooms of the private floor. A sitting-room for the Resident Physicians has been fitted up on the first floor, and adjoining it an office for the Chief Resident Physician. In the past year 1,894 cases were treated in the Hospital, 9,215 at the Dispensaries, 2,639 in the Receiving Ward; and 55 were treated at their homes by the Out-Patient Visiting Staff.

While the Managers desire to thank every donor for his gift, and a carefully prepared list of all donations, large and small, will be appended to this Report, special mention seems proper of several such donations.

Besides the $40,000 from the estate of Eliza W. S. P. Field, mentioned in last year's Report, the Hospital has received from the same source $5,000 for the John White Field Free Bed, and $5,000 for the Abby Willing Peters Free Bed. From Dr. George Fales Baker the Hospital has received $5,000, to endow a free bed in memory of his father, Alfred Gustavus Baker. $3,405.51 has been received for a children's free bed in memory of the late Dr. Edward Rhoads, collected from his many friends by Dr. James Tyson. The Hospital has to thank Dr. J. William White and Mr. George H. McFadden for tiling the Woman's Surgical Ward. A solarium for this ward has been built out of the funds furnished by the German-American Charity Ball. The Hospital is also indebted to Dr. J. William White for the donation of his X-ray apparatus, and it is now conducted as part of the Hospital property. The Athletic Association of the University gave to the Hospital a superheating hot-air apparatus. Joseph Zentmayer gave a first-class microscope, with an oil-immersion and two other lenses. One of the most useful gifts received by the Hospital was a new (Fulton & Walker) ambulance, presented by Mrs. T. A. Langstroth's Branch of the League of University Hospital Helpers. A small but complete operating room for the obstetrical wards has been built by the Auxiliary Committee, under the chairmanship of Mrs. G. Wharton Pepper. Dr. Henry Norris, Dr. David Houston, with other ex-Residents, papered and furnished the new sitting-room for Residents.

Dr. C. B. Penrose gave his instruments to the Hospital. To Mr. Henry F. Beale we are indebted for the use of his orchestra for one evening to give a concert for the benefit of the Hospital. The Hospital is indebted to Miss Baird (one of the Board of Women Visitors) for her very liberal donation of palms and other decorative plants for the wards and corridors, and also for her kindness in sending her gardener once a week to see that they are in healthy and thriving condition. To every contributor of small or large sums the Managers desire to say that every gift lightens their labors, and enables them more readily to conduct the Hospital. They feel also that they cannot state too strongly their obligations to the Board of Women Visitors, for the generous gifts of time and money in the betterment of the institution.

The Board hereby tenders to the Medical Staff of the Hospital and Dispensaries its appreciation and thanks for the skill, care and conscientiousness with which they have performed their duties.

Respectfully submitted,

FRANCIS P. SIBLEY,
Secretary.

APPENDIX VII.

REPORT OF THE DIRECTOR OF THE WISTAR INSTITUTE OF ANATOMY AND BIOLOGY.

To the Provost of the University;

SIR:—The progress of the Institute during the year has been satisfactory. Some of the results can be seen in the Museum, but much has been accomplished which is not apparent. The rebuilding of the Chicago properties belonging to the Institute, which I referred to in my last report, has occupied a large part of our attention, and under the personal supervision of the Assistant Director, is now nearing completion. This operation embraced, this year, the erection of a steam-heating plant and the rebuilding of twenty-six flats and five stores. The drawing of the plans, the purchase of materials and the direction of the work through several periods of labor strikes, has absorbed time which in past years has been devoted to the scientific work of the Museum. As, however, it has saved the Institute thousands of dollars, and as the work has been done so thoroughly that it will require little attention for years to come, the temporary check to activity in certain lines cannot be regretted. The funds of the Institute are in a highly satisfactory condition, and General Wistar is constantly adding to the endowment and devising methods of enlarging the future productive work of the Museum. The endowment consists of four special funds: two trust funds in the hands of distinct corporate trustees; the real estate owned by the Institute (not including the Museum building); and a large contingent fund which can be drawn upon temporarily for buildings and equipments, for repairs, or for such other large expenditure.

From a summary of the expenditures during the year,

we learn that the cost of repairs and current expenses of all kinds amounts to about 20¼ per cent of the total expense; permanent additions to the equipment to 15½ per cent; and salaries and wages 64¼. The cost of maintaining the institute building, including heat, light and current expenses, wages and administration, usually takes about one-third of our income; leaving two-thirds for equipment, purchase of specimens and current expenses of the Museum, and for salaries and wages of those engaged in scientific and Museum work. It should be remembered, however, that as the Institute enlarges its work, the cost of maintenance of the building will be proportionally much less.

During the past year the property has been kept carefully in repair, and some minor improvements effected. The walls of the exhibition rooms and hallways have been painted in oil, adding greatly to the attractiveness of the building. The heating and lighting plants continue to give satisfaction, and at a very low annual cost. The work of display in the Museum has been delayed for lack of proper cases. One more large iron and glass case has been completed. This was the first of a series of eight ordered from a reliable maker, but on its completion he found that he would be unable to furnish the rest at the stipulated price: hence it became necessary to find some other maker, or to do the work ourselves. As the plans of all the cases and of the special devices used thereon are our own, and can only be carried out under constant supervision, and as our experience in the construction of three iron cases enables us to estimate closely on the actual cost of production, it has been decided for the future to build the cases in our own shops. This decision has necessitated the purchase of heavier machinery, the rearrangement of the shop, the selection of skilled workmen, and the ordering of special material from the mills. All this has been done, and the first case is now under construction. The system of unit trays for the storage of disarticulated skeletons has been found satisfactory, and more sizes have been ordered. In a short time, therefore, all of this valuable material will be available for study. Owing to the want of proper jars to hold large dissections in alcohol, the work in Human Anatomy is still confined largely to the display of osteological preparations. During the year the Director has prepared the descriptive labels for the cases which will contain the remaining bones of the skull, and for many of them the specimens have been selected and made ready for mounting. The descriptions and display of only two bones of the skull already occupy five cases, hence some idea can be gained of the work required for the remaining twelve bones. That we are justified in spending so much time on this plan of text-book cases is shown, not only by the warm approval of experts, but by the large number of students who daily study the preparations. The most important addition to the exhibits in the Museum is a large group of orang-utans mounted naturally and artistically in the midst of foliage. All of the specimens were received unskinned, three immediately after death and three preserved in alcohol; hence it was impossible to get accurate measurements for mounting, and by means of death masks, to preserve the peculiar features of each animal. The group consists of one adult male, one adult female, two young males and two young females. Five of the specimens were presented by Dr. W. H. Furness. Of these, two died in the Philadelphia Zoölogical Gardens, two died on the passage to this country, and one was brought from Borneo in spirits. The sixth specimen was obtained from a dealer in living animals. It is believed that this group is the largest and most accurate in existence. It forms the first of the series of groups of the apes most nearly related to man.

An interesting group of fur seals, presented by the Biological Department of the University, has been mounted, and will be put on exhibition as soon as the case is completed. It comprises four specimens, a bull, a cow, a bachelor and a calf. The Harrison-Hiller collection of birds from Borneo have been mounted and temporarily displayed in a case in the lower hall. As soon as possible they will be labeled and arranged for exhibition in the Museum. The Institute has purchased the large McIlhenny collection of skeletons of Arctic mammals. It also contains a series of fishes and invertebrates, and a small but valuable lot of mammalian fetuses and viscera. Among the mammalian skeletons are one skeleton and ten skulls of the white whale; two skeletons and six skulls of the Pacific walrus; one skeleton and four skulls of the polar bear; three skeletons and twenty-three skulls of the ringed seal; two skeletons of the harbor seal; one skeleton of the bearded seal; two skeletons and eight skulls of the Arctic fox; three skeletons and fifteen skulls of the barren-ground caribou; and one skeleton and thirty skulls of the native Esquimo. We have also purchased the small Marudi collection from Borneo, containing several hundred mammals, reptiles, fishes and invertebrates. Through the kindness of Mr. Arthur E. Brown we have continued to receive valuable material from the Philadelphia Zoölogical Gardens, including, among other specimens, a polar bear, a camel, a deer, and two monkeys; and two Chinese dogs presented by Dr. Furness, with the understanding that they were to be sent to us when they died. The paleontological collection has received a few important additions.

During the summer Dr. Stotsenburg again used the Sea Isle Laboratory for the preparation of about fifty mammalian skeletons and a large number of skulls. An experiment in cleaning skeletons by insects was highly successful. Upwards of five hundred roughly cleaned and dried carcasses of animals of all kinds, varying in size from a bear to a mouse, were spread out on long tables in the basement rooms of the Institute. By the end of the summer almost the entire collection had been thoroughly cleaned by the larvæ of several species of beetles.

A number of scientific workers have used the laboratories during the past year, but none sufficiently long to call for detailed mention.

Respectfully submitted,

HORACE JAYNE,

Director.

APPENDIX VIII.

REPORT OF THE LABORATORY OF HYGIENE.

To the Provost of the University;

SIR:—I have the honor to submit the following report for the year ending August 31, 1900:

During the period named 352 students have received practical instruction in the Laboratory, as follows:

Undergraduates,	Medical	174
" "	Dental	157
Special students		21

The 331 undergraduate, and 19 of the special, students pursued studies in Bacteriology, while only 2 of the total of 352 applied for work in Practical Hygiene.

The didactic work comprised: A course of lectures on Bacteriology, two weekly, to the second-year Medical and the third-year Dental students. These lectures extended from the opening of the session to the Christmas recess. A course of lectures on General Hygiene, two weekly, to the Senior Medical students. These lectures extended from February 1 to the end of the session. Two courses of lectures on Ventilation and Heating, and on Water Supplies and Sewage Disposal, ten lectures in each course, to the students of Architecture in the College. Examinations were held on all of the above courses.

In addition to the teaching, we have, as hitherto, been called upon for miscellaneous investigations, analyses and examinations by outside individuals. In all, this work has amounted to a total of 96 chemical, bacteriological and microscopical analyses. Such time and energies as remained were devoted to problems coming properly within the scope of our work.

The Thomas A. Scott Fellowship was made vacant early in the year through the resignation of Miss Caroline Irwin, the appointee, and the electors did not deem it advisable to fill the position for the remainder of the term.

It is a pleasure to state that the new wing to our Laboratory, opened for occupancy and thoroughly equipped for practical teaching since the last report, has proven to be satisfactory in all important respects.

Since its completion, such a number of inquiries have been made concerning the details of construction, etc., that I deemed it advisable to prepare a careful description, with plans and photographs. As this was published in the *University Bulletin,* it is scarcely necessary to repeat it here.

It is equally a pleasure to say that the plan of concentration of teaching, advocated in preceding reports, has been adopted, and is meeting with the full approval of both students and instructors. In short, the experience of the past year, with both equipment and newly adopted methods, leaves no doubt that the usefulness of this Department to the University, in so far as the teaching of Bacteriology is concerned, is greater than ever before.

It is manifest, however, by a perusal of the statistics given at the opening of the report, that the demand for that side of our work more properly known as General Practical Hygiene has been growing annually less. At the time of writing (November 1, 1900), not a single student has applied for laboratory work in this field. I have searched diligently for the reason for this; and though I cannot say I have discovered it, I am still of the opinion that much could be done in the way of encouraging men to take up these studies were their efforts rewarded by some official recognition on the part of the University: such, for instance, as a degree or a certificate. Already certain other important teaching institutions are considering this plan, and I do not believe we can too promptly adopt it.

Preliminary steps toward the organization of a comprehensive course of study, for which a degree or certificate could properly be awarded, have already been taken by certain members of the several faculties; and it is hoped that in the near future a plan may be presented for which we may hope to secure your approval.

Respectfully submitted,

A. C. Abbott, M. D.,
Director.

APPENDIX IX.

REPORT OF THE DEAN OF THE DEPARTMENT OF DENTISTRY.

To the Provost of the University;

SIR:—I have the honor to submit to you the following report of the condition of the Department of Dentistry for the academic year 1899-1900:

The total number of students matriculated, 1899-00..		484
Of these, there were students of the third year	155	
Of these, there were students of the second year.....	159	
Of these, there were students of the first year	158	
Special students	12	
	——	484

Number of new matriculates, including those admitted to advanced standing		196
Of these, there were admitted upon presentation of diplomas and certificates	133	
Admitted upon examination	7	
Admitted to advanced standing	44	
Special students	12	
	——	196

Those admitted to advanced standing presented certificates from the following institutions:

Albany Medical College	1
Australian College of Dentistry, Melbourne, Australia....	1
Baltimore College of Dental Surgery	3
Brooklyn College of Pharmacy	1
Cincinnati College of Dental Surgery	1
College of Physicians and Surgeons, Edinburgh, Scotland..	2
Dental Hospital, London, England	1
Indiana Dental College	2
L'Ecole Dentaire, Paris, France	1

Marion Sims Dental College	1
Melbourne College of Pharmacy	1
New York College of Pharmacy	4
Northwestern University Dental School	1
Pennsylvania College of Dental Surgery	2
Philadelphia Dental College	3
Tufts College Dental School	1
University of Buffalo, Dental Department	7
University of Geneva, Switzerland	2
University of Heidelberg, Germany	1
University of Minnesota, Dental School	1
University of Pennsylvania, Medical Department	3
University of Southern California, Dental Department	1
University of Zaragoza, Spain	1
University of Zurich, Switzerland	1
Victoria University, Liverpool, England	1

The countries represented in the Department are as follows:

Middle States	286
New England States	69
Western States	28
Southern States	22
Pacific States	10
Africa	1
Argentine Republic	1
Australia	8
Canada	17
China	1
Cuba	1
Ecuador	2
England	4
France	4
Germany	7
Hawaii	1
Holland	1
Ireland	1
Mexico	1
New Zealand	7
Nicaragua	2
Peru	1
Portugal	1
Russia	1
Scotland	1

Spain	1
Sweden	1
Switzerland	3
United States of Colombia	1

SUMMARY.

United States	398
Foreign countries	86

The gradual elevation of the standard of entrance requirement for this Department, noted in my previous report, has operated to decrease somewhat the total number of matriculates for the session of 1898-99, as compared with the total number of matriculates for the preceding session—a result which was anticipated by the Faculty when the increase of entrance standard was decided upon. While the reduction in the total number of matriculates may be to some extent attributed to the more stringent entrance requirement, it is not chargeable to that factor alone, inasmuch as a considerable number of matriculates of the previous session failed to advance to higher classes at the end of the term, and were transferred to other institutions.

The decrease of twenty-nine matriculates from the total number of the previous session has, under existing circumstances, been an acceptable result, in that it reduces the total number of students to a point more nearly in harmony with the capacity of the Department, and renders more effective instruction possible.

It is as yet too soon to form correct conclusions as to the advantages which the increase of entrance standard has wrought, with respect to securing a better grade of student material as a foundation upon which to build a professional educational superstructure. It is undoubtedly true that the increased standard of entrance secures a body of students having more liberal general culture; but, as noted in my previous report, the differences in the character of high school instruction to some extent defeats one important purpose of the entrance standard, namely; that of selecting students whose education best adapts them for professional instruction. That this qualitative element is an important factor in any entrance requirement is shown by the fact that, notwithstanding the progressive increase in the entrance standard in this Department, its operation has in no degree diminished the number of failures to advance in course from one class to another.

The curriculum of the Freshman Year is at present overcrowded; a fact which causes some concern to the members of this Faculty, inasmuch as the average First-year student appears to be doing, and is required to perform, a larger amount of work than should reasonably be demanded. Under such circumstances it is evident that the best educational results cannot be attained. A readjustment appears to be necessary; and plans are under consideration for a better distribution of the work of the curriculum, which it is hoped will relieve the First-year students of the excess of work necessary under the present plan.

When a point has been reached where it may be deemed expedient to increase the length of the course of instruction by the addition of another year, opportunity will be afforded for the correction of difficulties inseparably connected with the present arrangement of time. The latter, as noted in my previous report, has been found by practical experience to be altogether too limited for the best professional training in Dentistry.

I am pleased to note the decided improvement in the character of the technical work which is being attained through the extension and elaboration of the laboratory method of instruction, systematized and supervised by the incumbents of the Chairs having these branches in charge. The improvement in technical results is most marked; but there has developed incidentally the need for a greater number of trained demonstrators to carry out properly the work of laboratory instruction. It is hoped that this need can be met by the appointment of suitable additional instructors so soon as they can be found, or trained from among those now under instruction in the Department.

A steady growth in the popularity of the Department

among the members of the dental profession in foreign countries is noticeable, and is practically evinced by the number of foreign students in attendance. These from foreign countries in attendance for the session 1899-1900 reached a total of eighty-six, as compared with a total of seventy-nine previously reported for the past session. No one factor is more largely responsible for this increase than the advanced standard of entrance requirements which has been adopted in the Department.

No changes in the administrative or teaching force of the Department have occurred during the session. The results of the year's work have been eminently satisfactory, and, taken as a whole, may be regarded as marking an advance upon that of previous years.

Respectfully submitted,

EDWARD C. KIRK,
Dean.

APPENDIX X.

REPORT OF THE DEAN OF THE DEPARTMENT OF VETERINARY MEDICINE.

To the Provost of the University;

SIR:—I have the honor to present to you the following report on the Department of Veterinary Medicine for the scholastic year 1899-1900:

There have been forty-six students in attendance during the year, of whom three were post-graduates. There have been no changes in the personnel of the teaching staff or in the curriculum. The work of the School has continued along the lines established during the preceding years. It is in many respects unfortunate that it has not been possible to add to and improve the course of instruction. The conditions and the character of veterinary work have changed materially during the past few years; and it is important that the training of students who wish to become veterinarians shall be such that they will be prepared as well as possible to meet the demands upon them in practical life. The need is increasing for better trained men in the veterinary profession. The value of veterinary work in the preservation and protection of the vast interests centred in the live-stock industry of the country, and the need for men competent to protect the public from diseases of animal origin, are now manifest. Moreover, there is a demand that is growing for veterinarians to teach in agricultural colleges, and to study in agricultural experiment stations, problems of importance to the live-stock raiser. To illustrate this demand, I may record the facts that Dr. John J. Repp, Dr. L. A. Klein and Dr. John H. McNeall, all graduates of this School, have recently been appointed to positions upon the Faculty of the State Agricultural College of Iowa. Dr. Repp is also Veterinarian to the Experiment Station of the same State. Dr. H. A. Meisner, another of our graduates, has, during the past year, been appointed State Veterinarian of Maryland, and Dr. S. H. McClure has been appointed Assistant State Veterinarian of Montana. Dr. John R. Mohler was a few months ago selected to fill the important position of Assistant Chief of the Pathological Division of the Bureau of Animal Industry of the United States Department of Agriculture.

These are the most recent, and but a few, of the appointments held by graduates of the Veterinary School. The School has not been able to furnish as many men to colleges, experiment stations, etc., as have been called for. Positions such as those referred to above, and public veterinary work, are important because the total value of the domestic animal is great, for the reason that animals are valuable *en masse*. On the other hand, private veterinary practice is in a flourishing condition, because the values of individual animals have increased. The live-stock industry was never more prosperous than now; and all kinds of animals are worth more, and justify greater care and protection, than ever before.

It is evident that there is a distinct need for thorough instruction in the veterinary sciences.

The Veterinary Department is meeting this need to a large extent, but falls short in some particulars. The shortcomings may fairly be ascribed to this Department's meagre income. It has never been possible to command the undivided time and energies of even a single member of the Veterinary Faculty. Instructing veterinary students, under existing conditions, must be secondary to teaching in other departments of the University, or to earning a livelihood outside of the University. The School has not, to a sufficient extent, had the best thought and efforts of its Faculty. Additional courses of instruction are needed which could be given well by teachers now attached to the School, if they could afford to take sufficient time from their other work.

It is encouraging to note that each year the students entering the Veterinary School have higher educational qualifications. More than one-half of the present classes are graduates of colleges or high schools.

Respectfully submitted,

LEONARD PEARSON,
Dean.

APPENDIX XI.

REPORT OF THE VETERINARY HOSPITAL.

To the Provost of the University;

SIR:—I have the honor to submit to you the following report of the work, condition and needs of the Veterinary Hospital, for the year ending August 31, 1900:

There were treated in the Hospital 1,291 animals, and in the free out-of-door clinics 2,913; an increase of 287 "in cases" and 426 "out cases" over the previous year. The following data from the records of the Hospital are of interest, as showing the rate of increase in the number of cases treated during each year since 1885:

	In cases.	Out cases.	Total.
Year ending August 31, 1886 . .	352	No record.	—
" " 1887 . .	430	"	—
" " 1888 . .	421	"	—
" " 1889 . .	497	"	—
" " 1890 . .	351	"	—
" " 1891 . .	419	1151	1570
" " 1892 . .	502	1323	1825
" " 1893 . .	641	1412	2053
" " 1894 . .	629	1601	2230
" " 1895 . .	814	1700	2514
" " 1896 . .	961	2013	2974
" " 1897 . .	958	2098	3056
" " 1898 . .	1081	2670	3751
" " 1899 . .	1004	2487	3491
" " 1900 . .	1291	2913	4204

The receipts from the stable, kennel and farriery for each year since 1890 have been as follows:

		Stable.	Kennel	Farriery.
Receipts from	1891 . . .	$2,194 11	$1,004 55	$1,328 12
"	1892 . . .	2,611 70	1,111 50	962 25
"	1893 . . .	*4,864 95	1,627 34	989 14
"	1894 . . .	*4,353 45	1,504 06	669 42
"	1895 . . .	3,155 70	2,037 08	814 00
"	1896 . . .	3,329 35	2,437 75	1,177 09
"	1897 . . .	3,237 90	2,438 45	1,209 53
"	1898 . . .	3,656 59	2,721 13	1,357 40
"	1899 . . .	2,983 60	2,575 93	1,404 35
"	1900 . . .	3,636 50	3,322 50	1,936 45

In the hospital for larger animals there has been a gratifying increase of business, though the capacity of this part of the Hospital has never been taxed. The accommodations are very good, equipment and service nearly all that is to be desired, and the fixed charges are exceptionally low. The slow growth of its business seems to be due rather to the large number of comfortable private stables in this city than to any shortcoming of the Hospital, or indifference of the community. The building is in a good state of repair, and the wards are bright, clean and comfortable.

The urgent needs are few. "Stocks" for securing fractious horses or mules in a standing position, while they are shod or subjected to minor operations, are of almost daily need. A frame shed, open on all sides for the reception of heat-stroke cases during the summer months, should be built in the centre of the grass plat in front of the stable.

The hospital for smaller animals has maintained a daily average of 20 dogs and cats. This is an excellent record, but it will undoubtedly be surpassed in 1901. The kennel building is in good repair, its equipment fair, and its accommodations still sufficient for present demands. A kitchen and a grooming room would add much to the efficiency of this branch of the Hospital. Runways should be built, into which dogs could be turned for short periods on clear days. The drainage from the cages should be modified in the three wards.

* Includes many uncollectable bills, which were subsequently "charged off."

The farriery has done the largest business in its history; but, owing to a necessary increase in wages, it is not yet self-supporting, though it bids fair soon to be. It is invaluable as an aid in the Hospital work and in the instruction of students. Its work consists largely in shoeing to cure lameness, and to correct irregularities of gait. Prices are moderate.

In general the work of the Hospital has been gratifying, and gives promise of permanence. The present system of management is good, but it seems as though it would be wise to employ a permanent Resident Surgeon at a moderate living salary, because under the present one-year tenure plan he no sooner learns the *clientèle* of the Hospital than he surrenders his place to his successor. The clients of the Hospital are brought into more frequent contact with the Resident Surgeon than with the clinical staff, and are no more pleased with having to deal with a strange Resident Surgeon each succeeding year than we should be were we compelled to change our family physician annually.

A clean, cheerful waiting-room for those who bring animals for treatment should be provided at once.

Respectfully submitted,

JOSEPH E. GILLINGHAM,
President.

APPENDIX XII.

REPORT OF THE LIBRARIAN.

To the Provost of the University;

SIR:—I have the honor to present the report of the Library for the year ending August 31, 1900.

During the year the work on the revision of the catalogue, referred to in my last year's report, was practically completed. In all, 139,963 volumes were re-catalogued. As a result of the undertaking, we now have a catalogue which I feel warranted in saying is both complete, inasmuch as it accurately represents every volume in the University collection; and useful, inasmuch as it enables readers not only to find what works of a certain author the Library possesses, but to ascertain quickly also what works on any given subject are represented in the Library's collections.

An important piece of work undertaken in connection with the revision of the catalogue was a careful comparison of the old cards with the books on the shelves, with the view of ascertaining what number had disappeared since 1884, when the old catalogue was begun. The number found missing was 914, though it is likely that by further search this number will be somewhat reduced. Fortunately the number of valuable works that have disappeared is small.

The result, however, of this examination demonstrates the necessity of making the most careful provisions for frequent examination of the shelves. The practice of University libraries differs somewhat in this respect. At some institutions the examination goes on constantly throughout the year; at others, the summer vacation is chosen as the time for such examination. During the year the time of our Library force is entirely taken up with the cataloguing of new books, and the experience of the last two years shows that during the summer vacation this work must be actively carried on. It may therefore be necessary, in case it is decided to carry on the examination every summer, to engage, for a few weeks, some additional help for the purpose. I may add that during the past two years only twelve books—seven in 1898-99 and five in 1899-1900—have disappeared from the Reference Library, and we have only noted nine as missing from the stack.

Accessions to the Library.

There have been added to the Library since September 1, 1899, 8,737 volumes. Of these, 3,960 were obtained by purchase, 564 through the Exchange Bureau, 663 were official documents of governments, States and municipalities, 2,971 through gift; while 579 volumes were received prior to the date mentioned, but not formally added to the catalogue until after the first of September, 1899. This number of accessions, it should be noted, includes only volumes received bound, or sent to the binder for the purpose of being bound. In addition, upward of 1,100 dissertations were received from European and American universities through the Exchange Bureau, besides a large number of minor government, State and municipal documents. While these are added to the catalogue cards, they are not accessioned until a sufficient number are obtained to make a volume.

The distribution of the books purchased from the various funds at the disposal of the Library is as follows:

	Volumes.	Cost.	Binding.
Comegys	11	$10 37	$3 20
Harrison	674	897 57	92 61
Krauth	80	118 08	7 48
Lippincott	114	258 77	24 03
Norris	181	532 14	44 80
Wagner	219	632 46	9 98
Pepper	132	410 97	15 76
Agriculture	12	16 93	1 55
Contributed funds	2,637	6,015 85	418 03
	4,060	$8,893 14	$617 44

The total amount thus spent was $9,510.58. The funds available during the past year, as will be seen from the above list, are: (1) The regular income of the Krauth, Lippincott, Norris, Wagner, Pepper and Agriculture funds; (2) the balances of the Comegys and Seybert funds, and of the State appropriation, as well as the balances of the special gifts referred to in my last year's report. The special funds received during the past year were: (*a*) A gift of $2,000 for current books distributed among twenty departments of the college; (*b*) a fund of $1,000 to defray the cost of subscriptions to periodicals; (*c*) the George L. Harrison Fellowship fund, amounting to $100 for each Fellow; (*d*) a gift of $1,000 from Mr. E. Burgess Warren for the purchase of books pertaining to the History of Statistics of American Commerce; (*e*) a fund of $17 for the purchase of books on the English Drama, contributed by students participating in the performance of the "Tamer Tamed" in the spring of 1899; (*f*) a fund of $100 from Mr. J. G. Rosengarten for the purchase of books on Colonial Administrations; (*g*) a contribution to defray the cost of English Government publications; and (*h*) the Hon. William Potter has, as in former years, paid the subscription of the Library for additional volumes of Hansard's "Parliamentary Debates." In addition to the above amount expended on books, $887.39 were spent in the Binding Department, making a total of $1,504.83 for binding; or a grand total for books and binding of $10,397.97.

Cataloguing Department.

In this Department 9,244 volumes were made ready for use, the cataloguing being distributed among the months of the year as follows:

	Works.	Volumes.	Cards.
Aug. 28 to Sept. 23, 1899..........	551	817	1,132
Sept. 25 to Oct. 28, 1899..........	518	675	1,195
Oct. 30 to Nov. 25, 1899..........	207	465	464
Nov. 27 to Dec. 23, 1899..........	265	434	740
Dec. 26 to Jan. 27, 1900..........	510	735	1,368

	Works.	Volumes	Cards.
Jan. 29 to Feb. 24, 1900..........	315	695	789
Feb. 26 to Mar. 24, 1900..........	723	946	1,610
Mar. 26 to Apr. 21, 1900..........	771	857	1,727
Apr. 23 to May 26, 1900	787	1,086	1,413
May 28 to June 23, 1900	751	1,011	1,555
June 25 to July 28, 1900..........	771	957	1,638
July 30 to Aug. 25, 1900..........	504	566	1,429
	6,664	9,244	15,060

In addition to this, the contents of 153 pamphlet boxes were catalogued. It will be observed that the total number of volumes catalogued exceeds the number accessioned, which is to be taken as an indication that during the year we were able to get rid of some of the accumulation of the past years.

In the Binding Department 4,256 books were forwarded to the binder, of which 1,779 were bound in cloth, 1,760 in boards, 61 in half-morocco and roan, and 656 repaired. During the year arrangements were made to have books purchased abroad bound before being sent to this country, so that for the future the work of the Binding Department will be somewhat diminished.

Early in the year the cataloguing of the Brinton Library was begun. The total number of books accessioned is now 4,098. In addition to this, the Library contains 183 MSS., 200 bound volumes not yet accessioned, and about 1,000 unbound pamphlets and magazines. In view of the character of this collection, which is almost wholly of an archæological nature, the Library Committee agreed to have the books deposited in the new Museum of Science and Art. A particularly valuable feature of this Library is the Berendt collection of MSS., a catalogue of which, prepared by the late Dr. Daniel G. Brinton, has been printed in the *Museum Bulletin* for May, 1900.

In July, the Provost placed at the disposal of the Library a special appropriation to defray the cost of cataloguing several collections received as gifts during the past years, but which it had not been found possible to catalogue through the medium of the regular force. A special force of cataloguers has been engaged for this purpose and is now at work. The total number of these books is about 10,000 volumes, and includes the valuable Macauley collection on Dante, Petrarch and Tasso.

In addition to this, we have on hand the library of the Philadelphia Society for Promoting Agriculture, consisting of about 700 volumes. It is proposed to use for the cataloguing of this collection the fund from the treasury of the Society which has been turned over to the University.

READING ROOM AND CIRCULATION DEPARTMENT.

In the use of the Library, both within the building and in the outside circulation of books, there has been a considerable increase during the past year.

The statistics of the inner Reading Room, used chiefly by the undergraduates of the College, by the students in the Graduate School, and in the Medical and Dental Departments, are as follows:

Oct. 2 to Oct. 28, 1899	2,161
Oct. 30 to Nov. 25, 1899	3,594
Nov. 27 to Dec. 23, 1899	3,340
Dec. 26 to Jan. 27, 1900	2,579
Jan. 29 to Feb. 24, 1900	1,778
Feb. 26 to Mar. 24, 1900	3,408
Mar. 26 to April 21, 1900	3,121
April 23 to May 26, 1900	3,674
May 28 to June 23, 1900	406
	24,061

By comparison with the statistics of last year, when the total amounted to 15,166, it will be seen that the increase is more than fifty per cent. During the year the greater facilities for the medical students in the reading room have resulted in a considerably increased attendance from this Department. The Pepper Medical Alcove is now filled with general reference works and modern text-books, and in addition to this the medical students have at their disposal the best current medical periodicals. We are indebted to Dr. Charles H. Frazier, who, as editor of the *University Medical Magazine,* forwards to us regularly the current numbers of thirty-nine important medical periodicals, so that we have now on file 129 of the leading medical periodicals.

The statistics of the Circulation Department are as follows:

The total number of readers registered at the delivery desk was 1,031, distributed as noted below:

Professors, Instructors and Officers of the University....		157
Students, divided as follows:		
The College	395	
Medical Department	129	
Dental Department	39	
Law Department	39	
Department of Philosophy	90	
Teachers' Courses	109	
Fellows	25	
		826
Readers paying $3.00 a year		4
Special readers		10
Employes of the University		34
		1,031

Number of books circulated during the year:

	Home use.	Library use.	Total.
Aug. 28 to Sept. 23, 1899	125	....	125
Sept. 25 to Oct. 28, 1899	1,680	2,433	4,113
Oct. 30 to Nov. 25, 1899	1,659	6,241	7,900
Nov. 27 to Dec. 23, 1899	1,870	5,819	7,689
Dec. 26 to Jan. 27, 1900	1,203	3,697	4,900
Jan. 29 to Feb. 24, 1900	1,511	3,320	4,831
Feb. 26 to Mar. 24, 1900	1,670	3,238	4,908
Mar. 26 to Apr. 21, 1900	1,340	4,088	5,428
Apr. 23 to May 26, 1900	1,517	4,723	6,240
May 28 to June 23, 1900	948	831	1,779
June 25 to July 28, 1900	637	203	840
July 30 to Aug. 25, 1900	351	163	514
	14,511	34,756	49,267

Circulation of books by classes:

	Home use.	Library use.
General Works	331	4,247
Philosophy	538	585
Religion	570	1,153
Sociology and Economics	2,271	8,638
Philology	500	674
Natural Sciences	613	794
Useful Arts	218	679
Medicine	831	2,444
Fine Arts	287	280
Literature	5,972	8,528
History	2,380	6,734
	14,511	34,756

Home Use	14,511
Library Use	34,756
	49,267

Number of books borrowed from other libraries:

Columbia University	10
Cornell University	2
Drexel Institute	2
Philadelphia Free Library	6
Hartford Historical Society	1
Harvard University	9
Philadelphia Library Company	27
	57

Institutions that have borrowed books:

Columbia University,
Johns Hopkins University,
Haverford College,
University of Chicago,
Swarthmore College,
Ursinus College,
Bryn Mawr College.

The circulation of last year amounted to 12,278 volumes for home use and 19,726 for library use,—a total of 32,004—showing an increase for this year of more than fifty per cent.

Departmental and Seminar Libraries.

During the year special catalogues in duplicate were prepared for books in the Departmental Libraries outside of the Library building, and for the Seminar Libraries in the Library building. In all, 6,124 cards were written for the former, and 2,175 cards for the latter.

The distribution of books in the Departmental and Seminar Libraries is as follows:

Departmental Libraries:

	Volumes	
Astronomy	400	
Botany	1,000	
Psychology	728	
Chemistry	1,092	
Hygiene	672	
Mechanical Engineering	640	
Civil Engineering	950	
Physics	181	
Architecture	482	
English	520	
Zoölogy	3,000	
		9,665

Seminar Libraries in upper floors of Library building:

Economics	574	
Ecclesiastical History	560	
English Economic History	520	
American History	125	
Philosophy	300	
Pedagogy	258	
Romanic Languages	1,166	
Latin	118	
Mathematics	450	
Germanic Languages	3,396	
		7,467
		17,132

For the English department it should be noted that the character of the books deposited is changed from year to year, in accordance with the courses offered by the instructors in charge. In the departments of Astronomy, Civil Engineering and Zoölogy there are also kept, besides the bound volumes, a large number of pamphlets not formally incorporated in the general Library; while the Germanic Seminar contains also a collection of about 5,000 dissertations in pamphlet boxes.

The year's experience has demonstrated the usefulness of the rooms in the upper floors of the Library, now set aside for part of the work done in the Graduate School. In all, twelve rooms have been fitted up. Of these, one is used exclusively as a seminar library for History and Economics; five are used as lecture rooms, and also contain libraries for the use of students within the respective departments; five are used as lecture rooms exclusively, and one as a test room for experiments in Psychology. The Economic and History Seminar is used more largely than any of the other special collections. During the day, as well as in the evening, it is generally well filled with graduate students engaged in researches of various kinds. The number of students using the other seminar libraries is smaller, but that is due in part to the smaller number of students taking the courses in question.

The rooms in the upper floors were formally opened for the use of the Graduate School early in December, and arrangements have been made to keep such as contain libraries open in the evening until eleven o'clock. In all these seminar libraries, with the exception of the Germanic collection, the professors themselves requested that the books should remain within the rooms; and this arrangement has worked very satisfactorily. The nature of the books, and the size of the collection kept in the Germanic Seminar, make it impracticable to carry out the rule in this instance; and a special attendant was engaged to be in the room at certain hours of the day, and hand out such books as students need for home use.

The study room fitted up in the hall for the benefit of undergraduate students has been found useful, and by means of it we are enabled to enforce more rigidly the very necessary law of silence in the Library, so that professors and students can be certain of securing absolute quiet.

Periodicals.

The total number of periodicals received regularly is 1,008, of which number 104 come to the Law Library and 904 to the main Library. Of this latter number, 236 come to us as gifts and 61 through the Exchange Bureau.

Law Library.

There have been added during the year—October 1, 1899, to October 1, 1900—1,532 books, making a total in the collection at present of 22,983.

The Law Library also receives as a donation the ordinances of twenty-three of the leading cities of the United States, and the publications of the Pennsylvania Bar Association and of the American Bar Association.

Number of Volumes in Library.

The number of books in the main Library, according to the accession book, is 167,058, and in the Law Library 22,938, making a total of 189,996. Adding to this the number now being catalogued—about 11,000—brings our total collection to 200,000 bound volumes and upward of 50,000 pamphlets. The latter, though unbound, are roughly classified and in a measure accessible.

Gifts.

The notable gifts during the past year were as follows:

The West Philadelphia Medical Book Club has generously turned over to the University its entire collection of medical periodicals and books. Many of the sets of periodicals duplicate our collection; and, with the consent of the Club, it is proposed to dispose of these by sale and use the money thus obtained for increasing the Medical Library.

The Bartram Memorial Library Committee presented

to the University its collection of botanical books which, while not large, contains some valuable publications, and makes a welcome addition to our botanical collection.

The medical publishers, Messrs. Lea Brothers, W. B. Saunders and P. Blakiston's Son & Co., of this city, have most generously, at the request of the Provost, forwarded their publications. Mr. Saunders has presented in this way upward of one hundred volumes, and Lea Brothers upward of seventy. Both of these publishers continue to send us their publications as issued.

Mr. J. Y. Jeanes contributed the amount required to purchase a very handsome set of Collier's Reprints.

Mr. Clayton McElroy has presented the papers and MSS. of the late Professor McElroy, embodying his lectures on Shakespeare, Chaucer and the English Language, delivered by him in the College during his incumbency of the Chair of English Literature.

Miss Esther White Harrison obtained for the Library a large collection of current books in French Literature.

The Hon. Mayer Sulzberger, of this city, presented an almost complete set of the American Jewish Publication Society's works.

A specially valuable gift was forwarded by President Diaz, of Mexico, consisting of an elaborate work on Mexican Antiquities in three folio volumes.

With a view to providing better library facilities for students of the Dental Department, an effort was begun last spring to obtain the leading dental periodicals. In response to requests sent to twenty-four publishers, nineteen responded favorably; and besides having on file in this way a considerable number of the best dental periodicals, Messrs. Claudius Ash & Sons, of London, and the S. S. White Dental Company, of this city, have sent us some of their recent dental publications. Most gratifying is the interest now being displayed by the students themselves to obtain a good dental collection. A committee has been appointed from among their number with the intention of raising a fund of $5,000, part of which is to be expended at once and the balance to be invested, the income only to be used for making additions to the Dental Library.

It is impossible to give a list of all who have kindly remembered the Library during the past year, but special thanks are due to Provost Harrison, Mr. William Darrach, Dr. Persifor Frazer, Mr. J. J. L. Houston, Professor W. D. Lewis, Dr. Caspar W. Miller, Dr. S. Weir Mitchell, Dr. J. H. Musser, Mr. Leonard Myers, Dr. Charles A. Oliver, Dr. George M. Rowe, Mr. J. G. Rosengarten, Mr. J. C. Sims, Dr. J. William White, Dr. Talcott Williams, Dr. H. C. Wood, who have from time to time forwarded miscellaneous collections of books, monographs and pamphlets. Among others who have sent copies of their own publications may be mentioned: Dr. H. V. Ames, Mr. Thomas Willing Balch, Dr. J. C. Essig, Professor A. Gudeman, Dr. H. A. Hare, Mr. Clarence B. Moore, Sir Joseph Sawyer, Dr. Alfred Stengel and Dr. Merrick Whitcomb. Dr. T. W. Koch, of the Class of 1892, has sent the University several hundred monographs and pamphlets on Dante, which form a most valuable addition to the Macauley collection.

In December, the gratifying announcement was made that the Academy of Natural Sciences of this city, and the heirs of the late Robert H. Lamborn, had consented to present to the University the library deposited by Dr. Lamborn a short time prior to his death. This collection, which consists of about 2,500 volumes, is exceedingly valuable in Travels and Archæology. In some respects the collection relating to Mexico is unique. The library is also distinguished by the large number of valuable and rare works which Dr. Lamborn obtained during a lifetime devoted to study and research. The Academy of Natural Sciences, which had the privilege of choosing such portions of the collection as it needed, generously waived this right.

Announcement has just been made of two gifts, one of $1,750, from the Class of 1889, to be spent in enriching our collection of philosophical books; the other, of $615, from Mr. Robert B. Buist, to enable us to purchase files of the most important botanical periodicals.

Of gifts other than books, the presentation of a bust of the late Dr. George B. Wood by members of the Wood family deserves to be recorded. On the last Alumni Day the formal presentation of the bust to the University was made in a fitting address by Dr. J. M. DaCosta (alas! no more with us), and the bust itself has now been placed in a position selected by the Library Committee in the main reading room. It stands on a pedestal, which records in an appropriate inscription the valuable services to the University and to medicine of one who well deserves to be remembered.

Finally, may I be permitted to point out, as last year, the needs of the Library? The total income from the invested funds is only about $1,500. We have been fortunate during the past years in securing from friends of the institution generous donations, by means of which we have been enabled to fill up some of the more important gaps in our collection, and also to provide for special work that is being carried on in various departments of the University. An improved catalogue and the facilities now provided in the Library have resulted, as the above statistics show, in a considerable increase in its use. It is safe to look forward to a further extension of such activity in the coming years; but in order to supply the demands made by the growth of the University in all directions, an assured income of $15,000 a year, to be expended in the purchase of books, does not represent more than a conservative estimate of what is necessary. For almost all branches of University work, a satisfactory basis has now been secured through the various collections obtained within the past ten years by purchase or gift, but in order to maintain the high grade of these collections it is obviously essential to add to them by steady purchases of new publications. The future usefulness of the Library will depend upon our ability to meet the growing needs of the various departments of the University.

Respectfully yours,

Morris Jastrow, Jr.,

Librarian.

APPENDIX XIII.

REPORT OF THE DEPARTMENT OF ARCHAEOLOGY AND PALEONTOLOGY.

To the Provost of the University;

SIR:—The year just elapsed has been one of development, and of adjustment to new conditions.

The principal care of the Department has been in testing the value of the changes made in its organization, and in placing its administration on an economic basis.

The opening of the Museum on Sundays from two to six was a new departure, made to meet an imperative public demand. The wisdom of the step has been demonstrated by the result. The average number of visitors who every Sunday avail themselves of the opportunity thus offered has been 272, while once recently the attendance reached 593. The daily number of visitors during the scholastic year, from January 1 to June 23, reached 19,621. Classes, schools and other educational bodies, have made use of the privileges offered to them to study the collections.

Last winter the Archæological Institute of America was granted the use of the Lecture Hall for its winter course of lectures. These, with the Wednesday popular lectures of the Department, at first drew small audiences. The latter have steadily increased, until now the lecture room is often crowded.

On April 19, 20, 21, the American Oriental Society held its annual meeting at the Museum, and much interest was manifested in the accumulation of scientific material then displayed.

In addition to the above scientific uses of the building, it has also been utilized for some of the University's social gatherings. The "Faculty Teas" were held there, and the interest which the collections aroused on those occasions among the University's guests was most gratifying.

At the close of the first year of its existence, the Museum may therefore be regarded as having fully performed the function planned for it by its founders. It has already become a centre, not only of archæological, but also of popular interest in the community. This is further demonstrated by the rapid increase in the subscribing membership of the Department. This, in 1899, was 405. At the present date, under the able management of the Committee on Membership, of which Mr. Robert C. H. Brock is chairman, it numbers 709.

Architecturally, the building has attracted much attention. It occupies a considerable space in the handsome volume for 1900 issued under the auspices of the Architectural League of America. A number of special inquiries with regard to details of the plant have been received from experts contemplating similar undertakings; and a copy of the plans was requested by a committee appointed by the Saxon Government with a view to the erection of a new Museum in Dresden. Dr. A. B. Meyer, Director of the Dresden Museum, with the well-known architect, Professor Wallot, also of Dresden, came to Philadelphia for the purpose of examining the building and gathering information for the commission.

In the autumn of 1899 a National Export Exposition was held under the auspices of the Philadelphia Museum; and application was made by its Board of Directors for the use of the unimproved land placed by the University under the charge of the Department. The application was granted; and in recognition of this courtesy the management of the Exposition, besides grading and otherwise improving the land, agreed to complete the brick and red granolithic pavement along the Spruce street front, from the Museum to South street bridge, at a cost of $1,800.

At the close of the Paris Exposition, the city of Manchester requested that the University's educational exhibit should be transferred to that city as a whole. The Department was asked to concur, and as no expense was involved, agreed to the proposal. The exhibit will not, therefore, be returned until the month of March, 1901.

The American Section has been enriched by a valuable collection of pottery, stone and other antiquities (chiefly from the vicinity of Natchez, Mississippi), formerly the property of Dr. Montroville W. Dickeson.

The most important accession is a series of objects from various tribes of Western Indians, collected by Mr. Stewart Culin during an expedition undertaken in the early part of the summer, under the patronage of the Hon. John Wanamaker. Sixteen tribes on twelve reservations from Iowa to Washington were visited, and some 2,000 specimens collected, many of which are of the greatest value.

The collection derived from the excavations of Dr. Uhle in 1896-97, at Pachacamac, Peru, has been arranged in the American Hall.

In the Section of Ethnology a notable gift has been made by Dr. W. H. Furness, 3rd, Dr. W. H. Hiller and Mr. Alfred C. Harrison, Jr. This represents the result of an expedition made by them in 1899 in the Naga Hills of Assam. The objects, similar to those obtained by the same collectors during their travels in Borneo, are now in course of installation. They have also presented a series of masks and other religious objects from Thibet, and three valuable fragments of Græco-Buddhistic sculpture. The special Ethnological collections have received numerous additions. Mrs. Frishmuth has made important accessions to her musical instruments, and Mrs. Joseph W. Drexel to her fans. Mrs. William H. Miller has given a valuable collection of American coins; and a large and important collection of Chinese, Japanese and Korean coins has been deposited by Mr. Edward W. Thwing. The entire collection of coins has been arranged for exhibition.

In the Babylonian and General Semitic Section, the campaign which has recently come to a close has been more important in scientific results than any undertaken since the beginning of the excavations. This was due to the presence in the field of Professor Hilprecht, who was sent to Nippur by the Babylonian Committee in February, and who, as scientific Director, joined Dr. Haynes and his party and supervised operations until May, 1900.

Besides numerous minor discoveries that will shed light upon the religious ideas and the daily life of the inhabitants of the lower Euphrates region, it was ascertained that the early limits and extent of Nippur were practically the same as they were in the Persian period. In tracing the city walls, the great Eastern gate, of pre-Sargonic origin, known from the late inscriptions, was discovered. The Temple Library was located, and partially excavated. Nineteen thousand inscribed tablets of unbaked clay were taken from the rooms and shelves, and in addition to these nearly five thousand more were gathered from different places in Nippur. Equal in importance with the discovery of the Library is that of an immense palace, six hundred feet long, built with two stories, and dating about 4000 B. C. The Ziggurrat of Ur-Gur had been determined to be but an enlargement of an earlier Sumerian stage-tower. The precise character and age of a building in the upper stratum of a mound, and known as the "Court of Columns," has been found to belong to the Parthian period; and it has been definitely ascertained that the upper stratum of the temple mound contains the remains of a huge fortress, sixty feet in height, of the same period. During the excavations over eleven hundred coffins were opened for the purpose of studying the modes of burial, and to gather data for determining the types of vases and other pottery in use at the different periods of Babylonian civilization.

The rich harvest of archæological material collected is too extensive and varied to enumerate here, but the liberality hitherto displayed by the Turkish Government towards Philadelphia encourages the hope that large and important accessions will shortly be added to the present series.

The work of the Egyptian Section has been conducted, as usual, in co-operation with the Egyptian Exploration Fund and Mr. W. M. Flinders-Petrie, the necessary funds being provided by the American Exploration Society.

Feeling that the agent employed by the French syndicate which for the past four years has conducted excavations at Abylos had inadequately done his work; and that the scientific importance of this site, linked as it is in Egyptian tradition with the Osirian myth, and the early beginnings of Egyptian history, warranted a more exhaustive examination; Mr. Petrie obtained from the Egyptian Government permission to work over the same ground. The season was therefore spent in excavating the tombs of six kings of the first dynasty: Kings Zet, Merneit, Den-Setui, Azab-Merbapa, Mersekha (Semempses) and Qa. The scientific results obtained have more than justified the effort. Serious progress has been made in the identification of these oldest monumental kings with the Manethonian list for the first dynasty, *i. e.*, with the Mena cycle; and in stripping the earliest pages of Egyptian history of the legendary mist by which not only ancient traditions, but also recent unsupported theory, had obscured them.

Through the liberality of the little group of earnest men and women who compose the American Exploration Society (foremost among whom are the Hon. Calvin Wells and Mrs. John Harrison), and to the generous treatment of the Society by the London Committee of the Egypt Exploration Fund, an important share of the ancient inscribed fragments which are now throwing such unexpected light upon the earliest stages of civilization have found their way to the University. Besides these fragments, and the great Stela of King Qa, are stone bowls and jar sealings inscribed in the names of these kings, and a series of the Stelæ of their domestics.

In addition to this material, the importance of which is difficult to overestimate, noteworthy accessions have been received: gold, jewels, amethyst, and other hard stone beads and amulets, and fine bronze objects of later times, c. 2500 B. C. Also a statuette of a man called Sepairi (XII Dyn.), of the best school of art of the period. Many other interesting objects from the same source are included in this year's results.

The Society's share of the great find of Egyptian Papyri made by Messrs. Grenfell and Hunt, working for the Egypt Exploration Fund, is shortly expected to arrive, and will be added to the collection of the Egyptian Section. This derives special interest from the fact that although the archæological material of the Section is of the greatest importance, it has hitherto been poor in MSS.

To the American Exploration Society the thanks of the Department are also due, for the gift of the five Etruscan sarcophagi of stone, carved in the effigy of the deceased; and for the lid of another of similar type, but of terra cotta. These sarcophagi, some of which still retain traces of painting, were discovered near Viterbo, in a Necropolis of the third century B. C., on the estate of Signor Piatti, who excavated them partly for the Society, and partly for the University of California. Twenty-two sarcophagi were recovered, sixteen of which were in a sufficiently good condition to warrant the expense of distant transportation. Of the eight which were secured for the American Exploration Society, one has been presented to the Peabody Museum of Harvard and one to the Boston Museum of Fine Arts.

The Pennsylvania Museum and School of Industrial Art has deposited with the Department a stone fragment of a scene of the adoration of the Sun-disk (reign of Akhen-Aten), obtained by Mrs. Jones Wister at Thebes.

The collections given by Professor Maxwell Sommerville have also been increased by notable accessions. Eight boxes have been received, which are not yet unpacked, but the contents of which will add to the interest of the important series already on exhibition.

It is also worthy of record that the superb mosaic pavement from Carthage, the gift of Mrs. Dillwyn Parrish, has been set in the floor of "William Pepper Hall."

The Library of Dr. Daniel G. Brinton, deposited with the Museum under an arrangement with the Trustees of the University, has been received and placed in the Elkins Library of the Museum. It is now being catalogued.

Two numbers of the *Museum Bulletin* have been published through the liberality of Mr. Robert C. H. Brock, and have been used in exchange for foreign and American publications.

The maintenance of such a plant is a heavy responsibility. In order to support it on its present footing, an annual expenditure of $20,000 may be regarded as a close estimate; but this does not include the expenses of field work or the purchase of objects. Those in charge entertain the hope that an institution, the broad educational importance of which has been so fully proved, will not be left unprovided for by the generous public; and that sufficient interest in its aims may be felt not only to secure the future of the existing plant, but as well to increase its scope and to develop its usefulness.

Respectfully submitted,

Sara Y. Stevenson,
Secretary.

APPENDIX XIV.

REPORT OF THE DIRECTOR OF THE ASTRONOMICAL OBSERVATORY.

To the Provost of the University;

SIR:—I have the honor to submit the following report of the condition of the Flower Observatory and the work carried on there during the past year.

The property of the Observatory calls for no comment beyond the statement that the buildings and instruments have been maintained in good condition, no material additions having been made.

With regard to the scientific work, the plan heretofore announced has been strictly followed. According to this plan the same lines of research are carried on from year to year, and as a result we have a very considerable accumulation of material which is believed to be of some scientific value. This plan does not, however, preclude the observers giving due attention to matters of special scientific interest when such are called for. Two such occasions arose during the past year, viz.; the meteoric display which was scheduled for last November, and the total solar eclipse of May 28.

The watch for meteors was kept up on the nights of the thirteenth to the sixteenth, inclusive, when clouds did not interfere. Most of the students in Astronomy at the University participated, the results were published in the *University Bulletin.* The total eclipse of May 28 was successfully observed by a party from the University stationed at Barnesville, Ga. The program adopted was as follows:

1. Observation of the time of the four contacts.
2. Visual examination of the inner corona with a four-inch telescope.

3. Determination of the wave length of the green corona line, formerly supposed identical with the 1474 line of Kirchoff's scale.

4. Sketches of the corona.

5. Observation, visual and photographic, of the shadow bands.

All of these undertakings were successfully carried out excepting 3 and the photographic part of 5.

The regular observational work accomplished during the year was as follows:

Zenith telescope observations for latitude...	2,103
Double stars measured with equatorial.....	434

Observation of minor planets as follows:

(256)	Walpurga	2
(308)	Polyxo	5
(324)	Bamberga	3
(336)	Lacodiera	5
(375)	1893 A L........................	6
(434)	Hungaria	2
	1899 E Y........................	7
		30

Since the installation of the Zenith telescope more than 7,000 latitude observations have been made. The results from October 1, 1896, to August 16, 1898, have been published as Part II, Vol. I, Publications of the University of Pennsylvania, Series in Astronomy. The second installment, bringing the work down to September 22, 1900, is well in hand and can soon be prepared for publication.

Of the double star measurements, 309 of those measured in previous years have been re-observed on from three to four nights each. 125 additional pairs have been measured, including some far more difficult than any heretofore observed with this instrument. Twenty-two new pairs have been discovered and measured. The manuscript giving results of measures of 922 pairs, from January 1, 1897, to October 15, 1900, is completed, and is ready for publication.

It is perhaps superfluous to add that a large part of the energies of the Director and his two assistants is devoted to the work of instruction. The courses given to both graduate and undergraduate students are believed to compare favorably with those offered by the best universities in this country. It is proposed in the future, as in the past, to give this work precedence over all other.

The library has received from various sources gifts of nine bound volumes and seventy-two pamphlets.

Respectfully submitted,

CHARLES L. DOOLITTLE.
Director.

APPENDIX XV.

REPORT OF THE DIRECTOR OF PHYSICAL EDUCATION.

To the Provost of the University;

SIR:—I have the honor of presenting the following report on the work of this Department during the scholastic year 1899-1900.

The total number of students examined was 449, distributed as follows: College 352, Law 19, Medical 49, Dental 25, Veterinary 4. As in previous years, most of the students from the professional schools came to me to obtain the necessary permit to become members of some team. The increased severity with which the University Athletic Committee has enforced the regulation requiring all students to undergo an examination before becoming eligible for the teams, has resulted in an increase in the number of permits granted, which this year amount to 246, as against 222 last year. The number issued to students in the College is 160, or 45.4 per cent of the number examined.

The re-equipping of the Gymnasium mentioned in my last report has resulted, as was hoped, in considerably augmented interest in gymnasium work. The attendance has been double that of last year; and in addition many students have been attracted to come in a few times to learn the various exercises, and then go over to the Houston Club to continue their work throughout the year—preferring the Club to the Gymnasium because of the facilities for bathing, so desirable after violent exertion.

There has also been established a gymnasium for women students, and an assistant director and instructor appointed. While various delays have occurred to prevent its full usefulness this year, considerable work has already been done in it; and by the commencement of the next college year I hope all will be in order to enforce practically the same system as with the male students.

The activities of Dr. Steele, the Students' Physician, have steadily grown, and show in a gratifying manner the confidence afforded him by the student body. This year he has treated professionally 352 students, necessitating 921 visits, of which 273 were house calls and 648 office. The distribution of his work is shown by the following table of students treated:

College	52
Philosophy	0
Law	9
Medical	114
Dental	167
Veterinary	10
	352

Of this number, twelve were affected with contagious diseases.

In the Students' Ward of the Hospital the total admissions last year were 51, while this year the number increased to 66, distributed as follows:

College	22
Philosophy	4
Law	5
Medical	24
Dental	11
Veterinary	0
	66

The total number of student days was about six hundred, as against eleven hundred for the preceding year, showing the average duration of illness to have been much less.

The more important results of Dr. Zimmerman's ophthalmological examinations are shown by the following table:

Total number examined	184
Number re-examined from previous class	8
Already wearing glasses	48
Color-blindness	1
Squint	4
Myopia (total)	18
Myopia (high)	6
Other organic disease	1

Of the 184 students examined, 41 were found whose condition required further treatment, and were advised accordingly.

Respectfully submitted,

CASPAR W. MILLER,
Director.

APPENDIX XVI.

REPORT OF THE DIRECTOR OF THE BOTANIC GARDEN.

To the Provost of the University;

SIR:—I am glad to be able to report that the past year has been one of steady improvement rather than of novel development. The most noteworthy change in the aspect of the Garden has been effected by a varied but judicious grouping of native rhododendrons. The collection formed part of a valuable gift received through the kindness of a trustee, Mr. Joseph S. Harris. The planting of these, and of other specimens collected with them by our gardeners, such as hemlocks, sumachs and thorns, has completed the covering of vegetation which we originally contemplated securing. The vistas now appearing, the varied contrast of foliage presented, and the generally pleasing effects of the grounds, have attracted a larger number of visitors than in any previous year.

A considerable share of the time of the Director and gardeners has been given to the identification and correct naming of the greenhouse collections, which now number 1,160 species and distinct varieties. Each specimen bears an aluminum label inscribed with the scientific and common names, the natural order to which it belongs, its native habitat, and where necessary, the name of the donor. The educational value of such a collection is thus greatly increased, for it truly constitutes a silent but continuous lecture. When the necessary funds are available, a like system will be extended to the specimens in the Garden which are only imperfectly labeled at present. A beginning has been made in setting up large boards in appropriate positions, on which tables of plant classification are arranged. These will greatly aid the student in grasping the affinities of the groups illustrated by the specimens around him.

Some valuable gifts and exchanges have been made during the year. Mrs. Conard, of Sharon Hill, presented eighty-three greenhouse and hothouse species, including twenty-one species of palm. Mrs. Durham sent from Santo Domingo a collection of orchids, tillandsias and succulent plants. Mr. Roberts LeBoutillier gave a group of the smaller cacti, native to the Southwestern States; also several orchids. The Messrs. Dreer presented specimens of the finer water lilies, while from Horticultural Hall and Washington Botanic Garden has come a variety of greenhouse duplicates.

The Director made a trip in April last with two members of the Botanical Society to the Cape Fear Peninsula, N. C., and secured about 370 plants typical of that locality. Thanks to the facilities we now possess for their culture, these have all made good growth. The total number of plants received during the year has been 620.

From the Garden numerous donations have been made of growing or cut specimens to the Girls' Normal and High Schools, as well as to several private schools of the city, to the Altoona High School, and to private individuals who have aided us in previous years.

The facilities of the Garden have been utilized by the students for research work, and at present six original papers are ready for publication. The materials for seven others are being gradually brought together, and I earnestly trust that some arrangement may soon be made for their publication.

The value of the institution as an educational centre is demonstrated by the constant benefits enjoyed by our students and by members of the Botanical Society of Pennsylvania and their friends, representing at least 230 individuals.

To all of these systematic instruction is given in the principles of plant life. I am glad, therefore, to be able again to record that a considerable share of the working expenses of the Garden has been contributed in the shape of fees and donations from the members. The Ladies' Auxiliary Committee, the formation of which I recorded in my last report, has faithfully and successfully continued its labors, and promises to become increasingly useful in bringing the general public into contact with our work.

Objection has at times been made by those who are unfamiliar with the true needs of the science, that the space of ground covered by the Botanic Garden is too valuable to be thus utilized. Over against such objection is the fact that the city is constantly aiming to secure and keep open similar areas throughout its bounds, and even in localities where land is more valuable. Our classes will succeed only if they form an integral part of a University education. For their proper conduct a small Botanic Garden such as that now in existence is indispensable. The relation in size and situation of the Garden to Biological Hall is perfect. If further proof of its appropriateness were necessary, such is furnished by recent action of London scientists. Though that city possesses magnificent public gardens on its outskirts, when the proposal was lately made to abolish Chelsea Garden and the Royal Botanic Garden—both of great value and in the heart of the city—the necessity of these as convenient centres for botanical instruction was recognized, and their preservation from the hand of the spoiler has been practically assured. I rejoice that you have fully sympathized with the needs of our department in this respect, and I would express the sincere hope that a like attitude may be preserved by our University Trustees toward the University Garden.

During the year some striking improvements on the Campus have been effected by the staff, in furtherance of Mrs. Harrison's plans for its embellishment. During March and April the grounds round the Hospital engaged attention; in May and June, the vacated section of Locust street, between Thirty-third and Thirty-fourth streets, was fundamentally altered, so that now it is one of the most attractive parts of the Campus. During the autumn it will be further improved by the planting of groups of shrubbery.

Respectfully submitted,

J. M. MACFARLANE,

Director.

APPENDIX XVII.

REPORT OF THE DIRECTOR OF THE UNIVERSITY VIVARIUM.

To the Provost of the University;

SIR:—I have the honor to submit the following report on the University Vivarium for the year ending August 31, 1900.

The great value of the Vivarium for both teaching and research has been abundantly demonstrated during the past year. It has been used daily by students and instructors for observation and experiment on living animals, and it has afforded a constant supply of fresh material for class work in Zoölogy. The advantage of having open pools and aquaria throughout the winter, and a supply of living animals constantly on hand for both teaching and investigation, cannot well be overestimated. Eight different subjects of investigation have been conducted in the Vivarium during the year, and a still larger number has been advanced by the facilities which it offers. Two volumes of Contributions from the Zoölogical Laboratory, consisting of papers published during the years 1898-99, were issued during the past year, and the volume for 1899-1900 is now nearly ready. The latter will contain about twenty different papers by the instructors and students of the Zoölogical Department. In this work of research the Vivarium is an increasingly important factor.

The marine aquaria, which at the time of my last report had not been started, are now in full operation. During the winter several attempts were made to keep animals in artificial sea-water, made from sea-salt and fresh water, but without any success. Finally, through the kindness of Peter Wright & Sons, a quantity of sea-water (about 3,000 gallons) was brought on one of their tugboats from about thirty miles at sea, and was transferred by means of tank wagons from the boat to the Vivarium. This water has been used continually since May 8, and the animals in these aquaria were never in better condition than at present. In fact several species of animals, particularly certain hydroids, crustacea and gasteropods, have reproduced abundantly, while the eggs of some other species have hatched normally when placed in these aquaria. To have animals breed in an aquarium is perhaps the best test of its satisfactory character, and in this respect I think our marine aquaria are unexcelled. The salt-water system in the Vivarium is what is known as a "circulating" one. The surplus water is stored in a dark cistern, where it purifies by bacterial action, and where the temperature is nearly uniform the year around. From this cistern it is lifted by compressed air into the aquaria, and at the same time it is well aërated. From the aquaria the water overflows through a sand and gravel filter into the cistern again. Thus the same water may be used over and over indefinitely. Our aquaria demonstrate the entire feasibility of maintaining, at small cost, marine aquaria at a distance from the sea; and the time cannot be far distant when many such aquaria will be established.

Strange as it may seem, the fresh-water aquaria have been more troublesome than the marine ones. This has been due principally to the impossibility of filtering sufficient quantities of the city water when it is at its worst. There is little doubt that a circulating system of our fresh-water aquaria would be more satisfactory than the present arrangement. The low temperature of the city water in winter and its high temperature in summer, together with its usual turbidity, renders it unsuitable for aquarium purposes. However, in spite of these difficulties, there has been very little mortality in any of our aquaria, and some of them have been quite satisfactory. Two different species of fresh-water fishes bred and reared their young in these aquaria during the past summer, and many forms of invertebrate animals have appeared fortuitously, and have established themselves in such numbers as to be of great value in class work.

The intense heat of the past summer was a severe test of the Vivarium; nevertheless the construction of the building is so satisfactory that the mortality among the animals was scarcely increased at all during the hottest periods. This is chiefly attributable to the fact that the building is a low, solid structure of brick and cement, and that the aquaria and animal rooms are sunk below the level of the ground, so that extremes of climate do not quickly affect the temperature within the building.

Our greatest need in connection with the Vivarium is a trained man who can devote all his time to the care of the animals. Carelessness or inattention on the part of an attendant may bring to naught the work of years. Already a year's work on hybridity, which was yielding valuable results, has been lost through this cause alone. We also need a good photographic outfit for use in the Vivarium. Many of our records can be accurately kept only by means of photographs, and a few efforts made with amateur outfits show that there are possibilities of very useful work in this line.

Finally, I wish to make suitable acknowledgments to Mr. Robert LeBoutillier for a hive of bees; to the A. I. Root Company for their publications on bee culture; to Mr. Arthur Rees for a monkey from the Philippine Islands; to Dr. Robert Formad for some valuable Chinese goldfish; and to various other persons for gifts of small animals.

Respectfully submitted,

E. G. Conklin.

APPENDIX XVIII.

REPORT OF THE DORMITORIES.

To the Provost of the University;

SIR:—I have the honor to present herewith my report on Dormitory receipts and expenditures for the fiscal year ending August 31, 1900.

EXPENDITURES.

Supplies	$204 64	
Care of buildings	4,816 85	
Heat and light	7,306 93	
Printing	193 50	
Repairs and additions	515 31	
Water rent	135 15	
Office expenses	524 04	
Goods for students	224 19	
		$13,920 61

RECEIPTS.

Rents	$27,476 98	
Charges	307 09	
		27,784 07
Balance		$13,863 46

Profit.—The net receipts for the year just closed show an increase of $950.43 over the net receipts for the previous year. The net receipts each year since the Dormitories were opened in the fall of 1896 show a steady increase, the figures for the four fiscal years being, respectively, $9,102.09, $12,554.18, $12,913.03, and $13,863.46. The reasons for the increase this year are to be found chiefly in the items of Care of Buildings, Supplies, and Repairs, in all of which there is a material reduction. These are discussed more fully below. The gross receipts from rentals and charges were, however, less than in the preceding year by $554.88, due to causes also presented below.

Heat and Light.—This item shows a total increase of $197.97. There is an increase of $260.36 in Heat alone, of $29.33 in Light, of $141.48 in Sinking Fund, and $58.13 in Bond Interest; and a decrease of $291.33 in the item of Special Charges. Per tenant, Heat and Light (including special charges, etc.) cost for the year $20.41, an increase of 39 cents over last year. Light alone cost per tenant $5.46, which is about the same as in 1898-99 ($5.42). I am disappointed in this. I had hoped that by the use of only eight candle-power lamps in the halls, etc., and by carefully watching the use of light, as we have done, we could materially reduce this expense. It is impossible, however, to control the use and waste of light in the rooms. The weather is also a difficult but constant factor, more light being used on dark days. We made last year a decided gain over the year before in the cost of light per tenant; it may be that we are down now to the limit under present conditions. It will be interesting to compare the costs with those of the new Dormitories, in which each room will have a meter.

The new coils put into the cellar in the summer of 1899 were the means of making comfortable last winter some rooms that have always been cold. Heat complaints were 183 in number during the year just closed and 227 the year before. Complaints as to the light-brackets, fuses, etc., were more numerous; and those as to imperfections in registers, etc., were also more numerous, which is probably to be expected, as these fixtures wear out with use. Though the complaints were less, the heating system is still not perfect, and many tenants were uncomfortable again. The system for the new Dormitories promises better results.

Goods for Students.—This, as before, includes window-shades supplied, floors stained, wood for fireplaces, etc. This account is less than last year's by $151.94, the reasons being the same as were shown in the report of a year ago.

Care of Buildings.—It is in this item that the most notable gain has been made. The cost of this service (which includes all care of rooms, halls, lavatories and grounds, and watching the buildings at night), has gone down steadily in the total and in the cost per tenant ever since the Dormitories were opened, but the percentage of saving is greater this year than ever. The net saving over last year is $1,233.98, and the saving per tenant is $3.08: the cost per tenant being only $13.46, as compared with $16.54. The cost of caring for 459 rooms, with the hallways and grounds attached to them, has been (per room) $10.50 for the year, or 87 cents a month. The cost per room a month last year was $1.09. This saving was anticipated in my report of a year ago, as when that was written we had already made changes in the service that I felt sure would result in economy.

Demand for Rooms.—The rentals for each year since the Dormitories were opened have been less than the total possible rental: that is, at no time have all the rooms been rented. The gross rental value of the Dormitories when opened was $28,185, which was increased to $28,800 in September, 1898, by the raising of the rents of some rooms. On the other hand, the gross rental receipts for the years ending August 31, 1897, 1898, 1899 and 1900 have been, respectively, $21,678.18, $26,795,39, $27,954.93 and $27,476.98. The reasons for this were quite fully stated in my last report: to what is said there I can only add that the causes operating against the rental of our double suites were acting even more strongly in 1899-1900 than in the previous year.

The demand for single rooms, on the other hand, was stronger than ever during the year just ended, and quite justifies the opening of the additional single rooms in the new buildings, all of which will, I think, be rented this year (unless they should be unfurnished when College opens).

Respectfully submitted,

E. W. Mumford,
Registrar.

APPENDIX XIX.

REPORT OF THE PUBLICATION COMMITTEE.

To the Provost of the University;

SIR:—The Publication Committee has continued its work during the past year as heretofore, and has been able to increase the efficiency of the University Publication Fund. The following matter has been published:

PHILOLOGY, LITERATURE AND ARCHÆOLOGY.

Vol. VII, No. 2.

The Rise of Formal Satire in England under Classical Influence. By Raymond Macdonald Alden. $1.00. Pp. 264.

UNIVERSITY BULLETIN (*Monthly*).

Vol. IV, Nos. 1-9.

Nine issues (two double numbers). Pp. 286.

ACCEPTED FOR PUBLICATION:

PHILOSOPHY.

No. 4.

Hindu Logic as Preserved in China and Japan. By Sadajira Sugiura.

POLITICAL ECONOMY AND PUBLIC LAW.

Railway Passenger Traffic. No. 16. By Walter E. Weyl.

HISTORY.

The Revolution of 1776 in Pennsylvania. No. 1. By Charles H. Lincoln.

Botany.

Seven Botanical Papers. Vol. II, No. 2. By John M. Macfarlane, *et al.*

Astronomy.

Description and Determination of the Longitude of the Flower Astronomical Observatory. Vol. I, Part I. By Charles L. Doolittle.

Foreign Exchanges.

Aachen, Technische Hochschule, Preussen, Deutschland.
Academie des Inscriptions et Belles Lettres, Institut de France, Paris, France.
Amsterdam-Koninklijke Akademie van Wetenschapen, Holland.
Amsterdam-Bibliothèque de l'Université (pour la Société Mathematique), Holland.
Asiatic Society of Bengal, Calcutta, India.
Basel, Universitäts-Bibliothek, Schweiz.
Berlin, Koenigl. Museum fur Naturkunde (Zoölogische Sammlung), Berlin, Deutschland.
Bern, Universitäts-Bibliothek, Schweiz.
Bernice Panahi Bishop Museum, Honolulu, Hawaii.
Bonn, Universitäts-Bibliothek, Preussen, Deutschland.
Braunschweig-Technische Hochschule, Deutschland.
Breslau, Koenigliche Universitäts-Bibliothek, Deutschland.
Brünn, K. K. Technische Hochschule, Mähren, Oesterreich.
Cambridge University Library, Cambridge, England.
Canadian Institute, Toronto, Canada.
Ceylon, India; Royal Botanic Gardens.
Constantinople, Musée Imperial Ottoman, Turkey.
Darmstadt, Technische Hochschule, Hessen, Deutschland.
Dorpat, Imper. Universitet, Dorpat, Russia.
Dresden Technische Hochschule, Sachsen, Deutschland.
Ecole Pratique des Hautes Etudes, Paris, France.

Erlangen, Koenigliche Universitäts-Bibliothek, Bayern, Deutschland.
Freiburg, i. B., Universitäts-Bibliothek, Baden, Deutschland.
Genève, Université de Genève, Suisse.
Genua, Reg. Universita degli Studi, Genua, Italia.
Giessen, Universitäts-Bibliothek, Deutschland.
Göttingen, Universitäts-Bibliothek, Preussen, Deutschland.
Göttingen, Koenigliche Gessellschaft der Wissenschaften, Deutschland.
Greifswald, Universitäts-Bibliothek, Preussen, Deutschland.
Halle, Universitäts-Bibliothek, Preussen, Deutschland.
Hanover, Technische Hochschule, Preussen, Deutschland.
Heidelberg, Universitäts-Bibliothek, Baden, Deutschland.
L'Institut Catholique de Paris, Paris, France.
Jamaica, Institute of, Kingston, Jamaica, W. I.
Japan, Department of Education, Tokyo, Japan.
Jena, Universitäts-Bibliothek, Sachsen-Weimar, Deutschland.
Kazan, l'Université Imperiel (de la Société Physics-Mathematique), Russie.
Karlsruhe, Technische Hochschule, Baden, Deutschland.
Kiel, Koenigl. Universitäts-Bibliothek, Preussen, Deutschland.
Koenigsberg, Koenigl. Universitäts-Bibliothek, Preussen, Deutschland.
Leiden, Rijks Universiteit, Holland.
Leipzig, Universitäts-Bibliothek, Sachsen, Deutschland.
Louvain, Université Catholique de, Belgique.
Lund, Kong. Universitet, Sverige.
Marburg, Universitäts-Bibliothek, Preussen, Deutschland.
McGill University Library, Montreal, Canada.
Moskau, Imper. Maskovskij Universitet, Moskau, Russia.
München, Technische Hochschule, Bayern, Deutschland.
Münster, Koenigl. Akademie-Bibliothek, Preussen, Deutschland.
Musée Social, Paris, France.
Nancy, l'Université de France.

New South Wales, Royal Society of, North Sydney, Australia.
Oxford University, Bodleian Library, Oxford, England.
Padua, Reg. Universita degli Studi, Padua, Italia.
Perugia, Reg. Universita degli Studi, Perugia, Italia.
Pisa, Reg. Universita degli Studi, Pisa, Italia.
Prag, Deutsche Universität, Boehmen, Oesterreich.
Prince Edward Island, Legislative Library of, Charlottetown, P. E. I.
Russian Journal, The, St. Petersburg, Russia.
St. Petersburg, Imper. Universitet, Russia.
Sapporo, Imperial College of Agriculture, Sapporo, Japan.
Strassburg, Universitäts-Bibliothek, Strassburg, Elsass, Deutschland.
Stuttgart, Technische Hochschule, Württemberg, Deutschland.
Toronto, University Library, Toronto, Canada.
Upsala, Kong. Universitets-Bibliotek, Upsala, Sverige.
Utrecht, Rijks Universiteit, Utrecht, Holland.
Wien, K. K. Universität, Oesterr.-Ungarn.
Wien, Technische Hochschule, Oesterr.-Ungarn.
Würzburg, K. K. Universität, Bayern, Deutschland.

Domestic Exchanges.

Academy of Natural Sciences Library, Philadelphia.
American Academy of Arts and Sciences, 1154 Boylston street, Boston, Mass.
American Academy of Medicine, Easton, Pa.
American Catholic Society, 715 Spruce street, Philadelphia.
American Jewish Historical Society, Washington, D. C.
American Philosophical Society, Philadelphia.
Andover Theological Seminary Library, Andover, Mass.
Brown University, Anatomical Museum, Providence, R. I.
Buffalo Historical Society, Buffalo, N. Y.
Buffalo Society of Natural Sciences, Buffalo, N. Y.
California Academy of Sciences, San Francisco, Cal.

California, University of, Berkeley, Cal.
Carnegie Library, Pittsburg, Pa.
Case Memorial Library, Hartford, Conn. (Hartford Theological Seminary.)
Catholic University, Washington, D. C.
Chicago Academy of Sciences, Lincoln Park, Chicago, Ill.
Chicago Theological Seminary, Chicago, Ill.
Chicago, University of, Chicago, Ill.
Cincinnati Society of Natural History, 312 Broadway, Cincinnati, O.
Colby University Library, Waterville, Me.
Columbia University Library, New York City.
Congress, Library of, Washington, D. C.
Cornell University Library, Ithaca, N. Y.
Denison University Library, Granville, Ohio.
Detroit Public Library, Detroit, Mich.
Free Library Bulletin, 1217 Chestnut street, Philadelphia.
Harvard University Library, Cambridge, Mass.
Illinois, University of (State Laboratory of Natural History), Urbana, Ill.
Indiana State Geologist, Indianapolis, Ind.
Industrialist, The, Kansas State Agricultural College, Manhattan, Kansas.
Iowa, University of, (Studies in Psychology), Iowa City, Iowa.
John Crerar Library, The, Chicago, Ill.
Johns Hopkins University Library, The, Baltimore, Md.
Kansas Historical Society, Topeka, Kansas.
Kansas University "Quarterly," Lawrence, Kansas.
Lebanon County Historical Society, Lebanon, Pa.
Leland Stanford University Library, Palo Alto, Cal.
Massachusetts Institute of Technology, Boston, Mass.
Michigan, University of, Ann Arbor, Mich.
Minneapolis Public Library, Minneapolis, Minn.
New Hampshire State Board of Health, Concord, N. H.
New Hampshire State Library, Concord, N. H.
New York, University of State of (Regent's Office), Albany, N. Y.

Northern Indiana Historical Society, South Bend, Ind.
Ohio Hospital for Epileptics, Gallipolis, Ohio.
Ohio State Arch. and Historical Society, Columbus, Ohio.
Peabody Museum, Harvard University, Cambridge, Mass.
Philadelphia, Library Company of, 1300 Locust street, Philadelphia.
Princeton University Library, Princeton, N. J.
Rush Medical College Library (University of Chicago), Chicago, Ill.
Smithsonian Institute, Washington, D. C.
Stanford University, California.
State Library, Harrisburg, Pa.
Tennessee, University of, "Record," Knoxville, Tenn.

Total receipts through Exchange Bureau:

564 printed volumes,
61 periodicals,
1200 Dissertations.

Respectfully submitted,

M. D. Learned,
Chairman.

APPENDIX XX.

UNIVERSITY BIBLIOGRAPHY,

Covering the period from September 1, 1899, to September 1, 1900.

ABBOTT, A. C.

The Epidemiology of Typhoid Fever. (*Proceedings Pathological Society of Philadelphia,* February, 1900.)

The Laboratory of Hygiene of the University of Pennsylvania: the Purposes for which it was Designed, and the Work being Done. (*Alumni Register,* February, 1900.)

A Description of the New Wing of the Laboratory of Hygiene, University of Pennsylvania. (*University Bulletin,* Vol. iv, No. 7, 1900.)

The Utilization of Bacteria and Bacteriological Methods in Sanitary Engineering. Address before the Engineers' Club. (*Proceedings of Engineers' Club,* Philadelphia, May, 1900.)

Some of the Developments and Demands of Modern Public Hygiene. Annual Address of the State Board of Health: delivered before the Associated Health Authorities and Sanitarians of Pennsylvania at Mechanicsburg, Pa., May 23, 1900. (*University Medical Magazine,* July, 1900.)

The Relative Infrequency of Acute Specific Infections in the First Year of Life. (*Transactions Association American Physicians,* 1900; *University Medical Magazine,* July, 1900.)

ADLER, LEWIS H., JR.

Some Practical Notes upon Diseases of the Rectum. (*New York Medical Journal,* December 30, 1899.)

Pruritus Ani, with Special Reference to Its Local Treatment. (*Philadelphia Medical Journal,* May 12, 1900.)

ALDEN, RAYMOND M.

The Rise of Formal Satire in England under Classical Influence. (Publications of the University, *Series in Philology*, etc., 1899.)

The Time Element in English Verse. (*Modern Language Notes*, December, 1899.)

Migration. (*Harvard Graduates' Magazine*, March, 1900.)

The Art of Debate. (Henry Holt & Co., New York, 1900, 279 pp.)

ALLEN, WILLIAM H.

Rural Sanitation in England. (*Yale Review*, February, 1900.)

The Rise of the National Board of Health. (*Annals of the American Academy*, January, 1900.)

ALLYN, HERMAN B.

Dilatation of the Heart. (*Transactions of the Philadelphia County Medical Society*, Philadelphia, November, 1899.)

AMES, HERMAN V.

Pennsylvania and the English Government, 1699-1704. (*The Pennsylvania Magazine of History and Biography*, April, 1900, Vol. xxiv, pp. 61-80.)

State Documents on Federal Relations: The States and the United States. No. 1: Interpretation of the Constitution during the First Two Decades of Its History. Edited with Notes, 44 pp. (Department of History, University of Pennsylvania, Philadelphia, 1900.)

State Documents on Federal Relations: The States and the United States. No. 2: State Rights and the War of 1812; 1809-1815. Edited with Notes, 44 pp. (Department of History, University of Pennsylvania, Philadelphia, 1900.)

ANSPACH, BROOKE M.

Tic Douloureux, Checked by the Salicylates. (*University Medical Magazine,* June, 1900.)

Neuritic Muscular Atrophy—not Hereditary. (*University Medical Magazine,* June, 1900.)

BATES, WILLIAM N.

Emendations to the Tenth Book of Pausanias. (*Proceedings of American Philological Association.* Vol. xxxi, 1900.)

The Coinage of the Ancient Greeks. (*Proceedings of the Antiquarian and Numismatic Society,* 1900.)

BERGEY, DAVID H.

The Necessity for Legislative Restriction of the Pollution of Surface Waters, through Sewage of Towns and Boroughs Draining into Ponds and Streams which Serve as Sources of Water Supplies for other Towns and Boroughs. (Report of the State Board of Health of Pennsylvania, 1898, pp. 388.)

Common Colds: their Cause, Prevention and Treatment. (*Philadelphia Medical Journal,* May 19, 1900.)

The Relation of Artificial Feeding to the Gastro-Intestinal Diseases of Infants. (*University Medical Magazine,* July, 1900.)

BOARDMAN, GEORGE DANA.

The Kingdom (Basileia). An Exegetical Study. (Charles Scribner's Sons, New York.)

BOHLEN, FRANCIS H.

Breach of One Instalment of a Divisible Contract, (*American Law Register,* July and August, 1900.)

BOLLES, ALBERT S.

Practical Banking. Tenth Edition, Revised. (Lerey Bros. & Co., Indianapolis, Ind., 1900.)

The National Bank Act, and Its Judicial Interpretation. Third Edition. (*American Banker,* N. Y., 1900.)

BRUMBAUGH, MARTIN G.

The Pennsylvania German. An Address before the Teachers' Institute of Berks County, Pa. (Pamphlet, pp. 8, *Reading Eagle.*)

Unknown and Yet Well Known. An Address to the Graduates of Juniata College. (Pamphlet, pp. 20, Rupert, Huntingdon, Pa.)

Educational Principles Applied to Literature. An Address to the National Educational Association, Charleston, S. C. (*Proceedings.*)

Thinking and Learning to Think. Pp. 350. By Dr. N. C. Schaeffer. Edited, and with an introduction, by M. G. B. (J. B. Lippincott Co., Philadelphia.)

Two Centuries of Pennsylvania History. Pp. 395. By President Isaac Sharpless. Edited, and with an introduction, by M. G. B. (J. B. Lippincott Co., Philadelphia.)

Record Book. Pp. 32. Prepared for Public Schools of Porto Rico. (*Boletin Mercantil,* San Juan.)

Course of Study. Pp. 48. Prepared for Public Schools of Porto Rico. (Public Printer, San Juan.)

CALVERT, PHILIP P.

A Contribution to Knowledge of the Odonata of Paraguay. (*Anales del Museo Nacional de Beunos Aires,* vii, pp. 25-35, 4 text figures, November 17, 1899.)

Order Odonata [in List of Insects of New Jersey]. (27th Annual Report of the *New Jersey State Board of Agriculture,* Supplement, pp. 66-75. Trenton, N. J., 1900.)

The Means of Defense of Animals. I. The Struggle for Existence. (*Scientific American Supplement,* New York, No. 1272, May 19, 1900, pp. 20,396-7.) II. The Protection of the Young, *l. c.,* No. 1276, June 16, pp. 20,456-7. III. The Protection of the Food Supply, *l. c.,* No. 1277, June 23, pp. 20,466-7. IV. Protection against Living Enemies [erroneously printed "animals" for "enemies"], *l. c.,* No. 1280, July 14, pp. 20,516-7. V. Protection against Climatic Changes, *l. c.,* No. 1281, July 21, pp. 20,535-7.

CHEYNEY, EDWARD P.

The Disappearance of Serfdom in England. (*English Historical Review,* January, 1900.)

Recent Writing on English History. (*International Monthly,* April, 1900.)

The Early Reformation in England. (*Translations and Reprints,* Vol. i, No. 1. Tenth Edition.)

CLARK, JOHN G.

Origin, Development and Degeneration of the Blood Vessels of the Human Ovary. (*Contributions to the Science of Medicine,* Baltimore, 1900.)

Historic and Practical Remarks on the Operative Treatment of Cancer of the Cervix. (*University Medical Magazine.*)

Treatment of Pelvic Inflammatory Disease by Non-Surgical Means. (*American Gynecological Journal,* April, 1899.)

CLARKE, HUGH A.

The Art of Vocal Harmony. (Silver, Burdett & Co., July, 1900.)

CONKLIN, EDWIN G.

Phenomena and Mechanism of Inheritance. (*Philadelphia Medical Journal,* September 16, 1899.)

Biology and Medicine: Biology as a Profession. (*University Bulletin,* November, 1899.)

The Fertilization of the Egg and Early Differentiation of The Embryo. (*University Medical Magazine,* March, 1900.)

CRYER, MATTHEW H.

Anatomical Variations of the Nasal Chamber and Associated Parts. (*Journal of the American Medical Association,* October 14, 1899.)

Some New Points of the Anatomy of the Face and Jaws. (*Dental Cosmos,* January, 1900.)

Prosthesis of the Jaw at the Time of Operation. (*University Medical Magazine,* June, 1900.)

Revision of Chapter on Extraction in Kirk's *American Text Book of Operative Dentistry.*)

CULIN, STEWART.

The Dickeson Collection of American Antiquities. (*Bulletin Free Museum of Science and Art,* University of Pennsylvania, Vol. ii, No. 3, January, 1900, pp. 113-168.)

The Origin of Ornament. (*Bulletin Free Museum of Science and Art,* University of Pennsylvania, Vol. ii, No. 4, May, 1900, pp. 235-242.)

Primitive American Art. (*University Bulletin,* April, 1900, pp. 7.)

Bibliography of Daniel G. Brinton, M. D., Memorial Volume. (American Philosophical Society, 1900.)

DAVIS, GWILYM G.

A Case of Double Congenital Luxation Hips: One Replaced by Manipulation, and the Other by Operation. (*Journal American Medical Association,* March 24, 1900.)

Method of Exposing the Gasserian Ganglion: Removal of the Superior Maxillary Nerve. (*Journal American Medical Association,* April 28, 1900.)

The Radical Cure of Hernia. (*Annals of Surgery,* May, 1900.)

Operative Treatment of Fractures of the Femur. (*Therapeutic Gazette,* May 15, 1900.)

Multilocular Cystic Growth of the Ovary. Accompanied by Increased Growth of Hair and Change of Voice, with Recurrence in the Abdominal Incision. (*American Journal of Obstetrics and Diseases of Women and Children,* Vol. xli, No. 5, 1900.)

Orthopædic Surgery. (Sajous' *Annual and Analytical Cyclopaedia of Practical Medicine.* Vol. v. F. A. Davis Co., Philadelphia, 1900.)

Perforation in Typhoid Fever from an Operative Standpoint, with a Report of Cases. (*University Medical Magazine,* May, 1900.)

DOOLITTLE, CHARLES L.

Results for Latitude and Aberration of Observations at the Flower Observatory, 1898-99. (*Astronomical Journal,* January 30, 1900.)

DUHRING, LOUIS A.

Dermatology and Syphilis. (*Year-Book of Medicine and Surgery,* 1899. Edited by George M. Gould, M. D. Wm. B. Saunders, Philadelphia, 1900.)

EASBY, WILLIAM, JR.

The Bacterial Treatment of Sewage in England. Paper read before the Engineers' Club of Philadelphia, May 19, 1900. (*Proceedings Engineers' Club of Philadelphia,* Vol. xvii, No. 3, July, 1900.)

EASTON, MORTON W.

Changes in Language. (*University Bulletin,* Vol. iv, No. 7, 1900.)

EDSALL, DAVID L.

The Influence of Immoderate Water Drinking upon Metabolism and Absorption. (*Contributions William Pepper Laboratory of Clinical Medicine,* 1900.)

Metabolism and Absorption During Exclusive Rectal Alimentation. (*University Medical Magazine,* March, 1900.)

A Critical Summary of the Literature on the Serum Diagnosis of Tuberculosis. (*American Journal of the Medical Sciences,* July, 1900.)

EVANS, HENRY BROWN.

Observations of Comets. 1899, III, and 1899, I. (*Astronomical Journal,* No. 482, May 29, 1900.)

FLEXNER, SIMON.

Experimental Pancreatitis. (*Contributions to the Science of Medicine.* Dedicated by his Pupils to William Henry Welch, 1900.)

Etiology of Tropical Dysentery. (*Philadelphia Medical Journal,* 1900.)

Report upon an Expediton sent by the Johns Hopkins University to Investigate the Diseases Prevalent in the Philippines. (*Johns Hopkins University Bulletin,* 1900.)

Medical Conditons Existing in the Philippines. (*International Clinics,* 1900.)

Actinomycosis of the Human Lung. (*Proceedings of the Pathological Society of Philadelphia,* 1900.)

Micro-organisms. (*Twentieth Century Practice of Medicine,* Vol. xix, 1900.)

FRAZIER, CHARLES H.

The Restitution of the Blood-plasma following Intravenous Saline Injections after Hemorrhage. (*Contributions William Pepper Laboratory of Clinical Medicine,* Philadelphia, 1900.)

An Experimental Study of the Etiology of Appendicitis. (*Contributions William Pepper Laboratory of Clinical Medicine,* Philadelphia, 1900.)

The Treatment of Gonorrheal Rheumatism. (*Therapeutic Gazette,* January 15, 1900, Detroit, Michigan.)

The Removal of an Appendix Filled with Oxyures Vermiculares. (*Proceedings of the Pathological Society of Philadelphia,* March, 1900.)

A Critical Summary of the Literature on the Surgery of the Stomach. (*American Journal of the Medical Sciences,* May, 1900.)

Osteoarthritis of the Spine and Hip Joints: Rhizomelic Spondylosis. (*University Medical Magazine,* July, 1900.)

FULLERTON, GEORGE S.

The Aim of the Teaching of Philosophy in American Colleges. Read at the Thirteenth Annual Convention of the Association of Colleges and Preparatory Schools of the Middle States and Maryland. (*Proceedings of the Association,* December, 1899.)

The Criterion of Sensation. Read before the American Psychological Association, New Haven, December, 1899. (*Psychological Review,* March, 1900.)

How Should Psychology be Taught? Delivered before the American Psychological Association, New Haven, December, 1899. (*Psychological Review,* March, 1900.)

GOODSPEED, ARTHUR W.

Seven Radiograms in Plates, III, IV, V, VI, VII, VIII, IX; illustrating article, "The Röntgen Rays in the Surgery of Children," by W. W. Keen, M. D. (*Cyclopaedia of the Diseases of Children,* Vol. v, Supplement. J. B. Lippincott Co.)

On the Technic of X-Ray Work. (*Philadelphia Medical Journal,* January, 1900, Vol. v, No. 1.)

Radiogram (Fig. 9) illustrating article, "The Use of the X-Ray in Orthopedic Surgery," by R. W. Lovett, M. D. (*Philadelphia Medical Journal,* Vol. v, No. 1, p. 69.)

Radiograms (Figs. 6 and 7) illustrating article, "A Case of Dilated Esophagus," by W. W. Keen, M. D. (*Philadelphia Medical Journal,* Vol. v, pp. 31, 32.)

Radiograms illustrating article on "Fractures." (*International Text-Book of Surgery,* Vol. 1, pp. 538, 545, 553, 584.)

Experience an Essential of Satisfactory Radiography. (*University Medical Magazine,* July, 1900.)

GRIFFITH J. P. CROZER.

Congenital Idiopathic Dilatation of the Colon. (*American Journal of the Medical Sciences,* September, 1899.)

Diagnosis in Congenital Diseases of the Heart; with Report of a Case. (*Philadelphia Medical Journal,* Septem- 30, 1899.)

Reports of a Series of Cases: Congenital Heart Disease with Great Cyanosis and Clubbing of the Fingers; Four Cases of Spastic Diplegia with Choreiform Movements; Stenosis of the Larynx in Typhoid Fever. (*Archives of Pediatrics,* May, 1900.)
Scarlatina Miliaris. (*Jacobi Festschrift,* 1900.)
The Diagnosis of Heart Disease in Children. (*Journal American Medical Association,* June 23, 1900.)

GUDEMAN, ALFRED.

The Achievements of Alexandria in Literature and Science. (*University Bulletin,* Vol. iv, No. 3-4, pp. 92-106.)
Agricola's Invasion of Ireland Once More. (*Classical Review,* xiv, pp. 51-53, 96.)
On Plutarch, Cicero, 5. (*Classical Review,* xiv, p. 62.)
Zu Petronius, c. 45. (*Berliner Philolog. Wochenschrift,* No. 6, p. 189.)
Zu Tac. Ann., i, 28. (*Berliner Philolog. Wochenschrift,* No. 10, p. 317 f.)
Note to Juvenal, i, 62; vii, 114. (*Classical Review,* xiv, pp. 158-160.)
Tacitus' Agricola and Germania: with Introductions and Notes, pp. lxxi, 295. (Allyn & Bacon, Boston, 1900.)

HAMILL, SAMUEL McC.

A Case of Nasal Diphtheria Complicated by Broncho-Pneumonia. (*Pediatrics,* Vol. x, No. 4.)
A Case of Migraine with Aphasia and Numbness in One Arm. (*University Medical Magazine,* August, 1900.)

HARSHBERGER, JOHN W.

Relation of State to Population. (*Education,* September and October, 1899.)
Philadelphia Botanists Commemorated in the Names of Plants. (*Plant World,* iii, 74.)
The Japanese Paulownia. (*Forest Leaves,* vii, 148.)
General Supervision and Revision of Glossary in "One Thousand American Fungi." McIlvaine, 1900, pp. 38. (The Bower-Merrill Co., Indianapolis, Ind.)

HENDRICKSON, WILLIAM F.

Musculature of the Duodenal Portion of the Common Bile-duct and of the Sphincter. (*Anatomischer Anzeiger,* xvii Band, Nos. 10 and 11, 1900.)

HIRST, BARTON COOKE.

Text-Book of Obstetrics. Second Edition. (Wm. B. Saunders, Philadelphia, November, 1899.)

A New Operation for Complete Inversion of the Uterus. (*American Journal of Obstetrics,* November, 1899.)

Coccygodynia. (*University Medical Magazine,* April, 1900.)

Some Recent Experiences in Intestinal Surgery. (*American Journal of Obstetrics,* Vol. 41, No. 5, 1900.)

HOLMES, EDMUND W.

The Deep Fascia. (*Proceedings Association American Anatomists,* December, 1899.)

Typhoid and Our Water Supply. (*Pennsylvania Medical Journal,* January, 1900.)

Anatomy of Hanging. (*Proceedings Philadelphia County Medical Society,* March, 1900.)

Medical Education. (*Pennsylvania Medical Journal,* September, 1900.)

JASTROW, MORRIS, JR.

Canaan. (*Encyclopaedia Biblica,* Vol. i.)

The Historical Study of Religions in Universities and Colleges. (*Journal of the American Oriental Society,* Vol. xx.)

The Name of Samuel and the Stem "Sha'al." (*Journal of Biblical Literature,* Vol. xix.)

The First International Congress of the History of Religions. (*International Journal of Ethics,* Vol. x, No. 4.)

The Tearing of Garments as a Symbol of Mourning. (*Journal of the American Oriental Society,* Vol. xxi.)

The Fourteenth Chapter of Genesis, and Recent Researches. (*Jewish Quarterly Review,* Vol. xiii, No. 49).

JOHNSON, EMORY R.

The Principles of Governmental Regulation of Railways. (*Political Science Quarterly,* March, 1900.)

KELLY, ALOYSIUS O. J.

Observations on the Gruber-Widal Reaction in Typhoid Fever. With Dr. A. A. Uhle. (*Philadelphia Medical Journal,* March 3, 1900.)

Tumors of the Vermiform Appendix. (*Transactions of the Pathological Society of Philadelphia,* March, 1900; *University Medical Magazine,* May, 1900.)

The Pathology of Appendicitis, in "A Treatise on Appendicitis," by Dr. John B. Deaver, Second Edition. (P. Blakiston's Son & Co., 1900.)

Papillomatous Epithelioma of the Pelvis of the Kidney. (*Transactions of the Pathological Society of Philadelphia,* July, 1900.)

KOENIG, AUGUSTUS O.

Revision of Essig's Dental Metallurgy, Fourth Edition. (Lea Brothers & Co., Philadelphia, 1900.)

LAMBERTON, WILLIAM A.

The Mutual Relations of Colleges and Schools. Read before State Teachers' Association of Pennsylvania, July, 1900. (*Proceedings.*)

LEARNED, MARION D.

Professor George Allison Hench: In Memoriam. (*Americana Germanica,* iii, 2, 219 ff.)

The "Lehrerbund" and the Teachers of German in America. (*Pädagogische Monatsheften,* I, 1, 10 ff.)

Volkserziehung und Weltkultur. (*Pädagogische Monatsheften,* i, 9.)

Leiter der Abteilung für höheres Schulwesen in der *Pädagogischen Monatsheften.*

LEWIS, WILLIAM DRAPER.

A Digest of Decisions and Encyclopædia of Pennsylvania Law, 1754-1898. Vols. vi, vii, viii. Co-editor with George Wharton Pepper. (Philadelphia: Rees Welsh & Co., 1899.)

World's Best Essays. Vol. i. (St. Louis, Mo.: Fred. P. Kaiser, 1900. Member of the Advisory Council.)

Address Accepting the New Law School Building on Behalf of the Faculty of the Department of Law. (*American Law Register and Review,* March, 1900. Vol. 48, No. 3.)

LINDSAY, SAMUEL McCUNE.

The Liberal Culture Element in Commercial Education. Address before International Commercial Congress, October 20, 1899. (*Proceedings.* Philadelphia, 1899.)

The Philadelphia Negro. Introduction to volume prepared by Dr. Wm. E. B. DuBois and Miss Isabel Eaton. (Publications of the University, *Series in Political Economy and Public Law,* No. 14. Philadelphia, 1899.)

Notes on Bibliography of the Negro Question. Published in "Race Problems of the South." (*Proceedings of First Annual Conference* at Montgomery, Ala. Richmond, Va., 1900.)

LINGELBACH, WILLIAM E.

The Doctrine and Practice of Intervention in Europe. (*Annals of the American Academy,* July, 1900.)

LLOYD, MORTON GITHENS.

The Transversal Thermomagnetic Effect in Bismuth. (Published, Philadelphia, June, 1900, by the Author.)

McMASTER, JOHN BACH.

Daniel Webster: A Study for the *Century Magazine.* Four papers. (The first of the series in the *Century Magazine* for November, 1900.)

MERRICK, J. HARTLEY.

Some Historical Notes on the Seal of the University. (*University Bulletin,* March, 1900. 12 pp. Illustrated.)

MIKELL, WILLIAM E.

Outline of a Course of Lectures on the Common Law of England. (Brandt Press, Trenton, N. J., 1899.)

MILLS, CHARLES K.

The Hospital and the Nurse. (*Philadelphia Monthly Medical Journal,* November, 1899.)

Anomia and Paranomia; with Some Considerations Regarding a Naming Centre in the Temporal Lobe. (*Journal of Nervous and Mental Diseases,* December, 1899.)

A Case of Unilateral Progressive Ascending Paralysis, Probably Representing a New Form of Degenerative Disease. (*Journal of Nervous and Mental Diseases,* April, 1900.)

Tumor of the Superior Parietal Convolution, Accurately Localized and Removed by Operation. With W. W. Keen, M. D.: Pathological Report by Wm. G. Spiller, M. D. (*Journal of Nervous and Mental Diseases,* May, 1900.)

Some Neurological Medico-Legal Experiences and Reflections. (*Colorado Medical Journal,* September, 1900.)

List of Publications of Charles K. Mills, M. D., 1872-1899. Philadelphia, 1900. (Printed privately.)

MOORE, J. PERCY.

A Description of Microbdella biannulata, with Special Regard to the Constitution of the Leech Somite. (*Proceedings of the Academy of Natural Sciences of Philadelphia,* 1900, pp. 50-73, Plate VI.)

Note on Oka's Biannulate Leech. (*Zoologischer Anzeiger,* xxiii, No. 623.)

MUIR, E. STANTON.

The Action of Certain Somnifacients on the Horse. (*Journal of Comparative Medicine and Veterinary Archives,* April and May, 1900.)

MUNRO, DANA C.

Selections from the Laws of Charles the Great. (Translations and Reprints from the Original Sources of European History, Vol. vi, No. 5. Department of History, University of Pennsylvania.)

MUSSER, JOHN H.

A Practical Treatise on Medical Diagnosis. Third Edition. (Lea Brothers & Co., 1899.)

Cancer of the Common Bile Duct. (*University Medical Magazine,* September, 1899.)

Hemorrhagic Diathesis in Typhoid Fever. With Dr. Joseph Sailer. (*International Medical Magazine,* November, 1899.)

Meralgia Paresthetica. With Dr. Joseph Sailer. (*Journal of Nervous and Mental Diseases,* January, 1900.)

Some Cases of Dilatation of the Stomach. With Dr. J. D. Steele. (*American Journal of Medical Sciences,* February, 1900.)

On the Uses of Antitoxin in Diphtheria. (*University Medical Magazine,* March, 1900.)

The Indications for the Use of Alcoholic Stimulants in Typhoid Fever. (*Therapeutic Gazette,* April, 1900.)

Diseases of the Liver. (Keating's Cyclopædia, New Edition by Edwards, 1899.)

Angina Pectoris. (Gould's *Dictionary of Medicine,* 1899.)

PACKARD, FRANCIS R.

Innocently Acquired Syphilitic Infection of the Throat. (*Journal of Eye, Ear and Throat Diseases,* Baltimore, Md., October, 1899.)

Traumatic Perforations of the Membrana Tympani. (*Transactions Philadelphia County Medical Society,* December, 1899.)

The Earliest Recorded Autopsies in America. (*Philadelphia Medical Journal,* February 17, 1900.)

Some Old Certificates of Proficiency in Medicine. (*Philadelphia Medical Journal,* March 10, 1900.)

PEPPER, GEORGE WHARTON.

A Digest of Decisions and Encyclopedia of Pennsylvania Law, 1754-1898. Vols. vi, vii, viii. With William Draper Lewis, Ph. D. (Philadelphia: Rees Welsh & Co.)

POTTS, CHARLES S.

Two Cases of Pseudo-muscular Hypertrophy Occurring in Brothers. (*Journal of Nervous and Mental Diseases,* February, 1900.)

A Case of Injury to the Spinal Cord, with Loss of Knee Jerk and Presence of Ankle Clonus; followed by Dissociation of Sensation of Hysterical Origin in Nearly One-Half the Body. (*Journal of Nervous and Mental Diseases,* February, 1900.)

A Manual of Nervous and Mental Diseases. (Lea Bros. & Co., February, 1900.)

QUINN, ARTHUR H.

The Last Five Yards. (*Saturday Evening Post,* Philadelphia, October 28, 1899.)

Pennsylvania Stories. (Penn Publishing Company, Philadelphia, November, 1899.)

George Eliot's Silas Marner. Edited with an Introduction and Notes. (Globe School Book Company, New York, August, 1900.)

RANDALL, B. ALEXANDER.

Four Operations for Cerebellar Abscess: One Success, Two Autopsies. (*Journal American Medical Association,* p. 1185, November, 1899.)

The Stacke Operation of Simple Exenteration of the Tympanic Cavities. (*Transactions American Otological Society*, p. 169, 1899.)

Lateral Sinus Phlebitis After Otitis Media in Typhoid Fever. (*University Medical Magazine*, December, 1899.)

RAVENEL, MAZŸCK P.

The Making of Agar-Agar. (Read at a meeting of the American Public Health Association, October 31, 1899. (*Reports and Papers of Association*, Vol. xxv, p. 605.)

Experimental Inoculation of Calves with Syphilitic Virus. (*Pathological Society of Philadelphia*, February 8, 1900; *Proceedings of the Society*, New Series, Vol. iii, No. 5; also *American Journal of the Medical Sciences*, April, 1900, and *Journal Comparative Medicine and Veterinary Archives*, May, 1900.)

Experimental Tuberculosis in a Goat. With Dr. C. Y. White. (*Pathological Society of Philadelphia*, February 8, 1900; *Proceedings of the Society*, New Series, Vol. iii, No. 6.)

Passage of the Bacillus of Koch into the Milk of a Tuberculous Woman. Translated. (*Journal of Comparative Medicine and Veterinary Archives*, June, 1900.)

A Case of Aneurism in a Chicken. With Dr. Joseph Sailer. (*Pathological Society of Philadelphia*, March 22, 1900; *Proceedings of the Society*, New Series, Vol. iii, No. 7; also *Journal of Comparative Medicine and Veterinary Archives*, April, 1900.)

Rabiform Symptoms due to Spiroptera. Translation. (*Journal of Comparative Medicine and Veterinary Archives*, June, 1900.)

Three Cases of Tuberculosis of the Skin due to Inoculation with the Bovine Tubercle Bacillus. (*Pathological Society of Philadelphia*, June 14, 1900; *Proceedings of Society*, New Series, Vol. iii, No. 10; *Philadelphia Medical Journal*, July 21, 1900.)

A Case of Pneumonomycosis due to the Aspergillus Fumigatus. With Dr. Leonard Pearson. (*Pathological Society of Philadelphia,* May 24, 1900; *Proceedings of Society,* New Series, Vol. iii, No. 10; also *University Medical Magazine,* August, 1900, and *Journal of Comparative Medicine and Veterinary Archives,* August, 1900.)

Dourine and its Parasite. Translation. (*Journal of Comparative Medicine and Veterinary Archives,* July, 1900.)

The Rapid Diagnosis of Rabies. With Dr. D. J. McCarthy. (*Pathological Society of Philadelphia,* June 14, 1900; *Proceedings of Society,* Vol. iii, No. 9; also *Journal of Comparative Medicine and Veterinary Archives,* July, 1900.)

Mumps in the Dog. Translation. (*Journal of Comparative Medicine and Veterinary Archives,* August, 1900.)

RENNERT, HUGO A.

Gregorio Sylvestre and his *Residencia de Amor.* (*Modern Language Notes,* November, 1899.)

Macias O. Namorado; a Galician Trobador. (Philadelphia, 1900, 64 pp. Privately printed.)

RHEIN, MEYER L.

Successful Prosthesis After Resection of Lower Maxilla. (*Items of Interest,* September, 1899.)

Aseptic Environment. (*Items of Interest,* November, 1899.)

Should the Medical Undergraduate be Instructed in the Principles of Dentistry? Read before the American Medical Association at Atlantic City, June 5-8, 1900. (*Journal of the American Medical Association,* June 23, 1900.)

An article prepared to open the discussion on Dr. H. W. Pierson's paper on "Constitutional Effects of Amalgam Fillings and Red Vulcanite Plates upon the Human Organism." Read before the Chicago Dental Society. (*Dental Review,* May, 1900.)

Prosthetic Dentistry. (*Dental Cosmos,* November, 1899.)
Pyorrhea Alveolaris. (*Ibid.,* October, 1899.)
The Possibility of Jumping the Bite. (*Ibid.,* February, 1900.)
Alveolar Necrosis. (*Ibid.,* May, 1900.)
Pyorrhea Alveolaris. (*Ibid.,* May, 1900.)
Education. (*The Journal of the American Medical Association,* July, 1900.)
Interstitial Gingivitis. (*Ibid.*)

ROTHROCK, JOSEPH T.

Forestry in Pennsylvania. 20 pp. Annual Report in Report of Secretary of Agriculture for 1899. (Harrisburg, Pa.: State Printer.)

SCHAEFFER, NATHAN C.

The School in Its Relation to the Higher Life. An Address before the National Educational Association. (*Proceedings* for 1899.)
The Annual Report of the Department of Public Instruction of Pennsylvania, for the School Year 1898-'99. (December, 1899.)
An Introduction to the Life of Rev. Henry Harbaugh, D. D. (Published by the Reformed Church Publication Board, pp. 19, Philadelphia, 1899.)
The Worth of Woman as a Teacher. An Address before the Convention of City and Borough Superintendents. (*Proceedings;* also *Pennsylvania School Journal,* March, 1900.)
Teachers' Certificates. An Address before the State Teachers' Association. (*Proceedings;* also *Pennsylvania School Journal,* August, 1900.)
Thinking and Learning to Think. Vol. 1, pp. 351, of Educational Series. Edited by Dr. Martin G. Brumbaugh. (J. B. Lippincott Co., Philadelphia, 1900.)

SCHELLING, FELIX E.

Ben Jonson and the Classical School. (*Publications of the Modern Language Association*, 1898.)

A Book of Seventeenth Century Lyrics. Edited with Introduction and Notes. (Ginn & Co., 1899.)

Tom Tyler and His Wife; an Excellent Old Play. Reprinted with prefatory words from the edition of 1661. (*Publications of the Modern Language Association*, 1900.)

SCHWATT, ISAAC J.

Secondary Algebra. With Professor G. E. Fisher. A text-book: pp. 397, xviii. (Published privately, 1900.)

Rudiments of Algebra. With Professor G. E. Fisher. A text-book: pp. 183. (Published privately, 1900.)

SMITH, EDGAR F.

Observations on Aconitic Acid. (*Journal American Chemical Society*, 1899.)

Action of Hypophosphorous Acid upon Molybdic Acid. (*Ibid.*)

Action of Hydrochloric Acid Gas upon Sulphates, Selenates, etc. (*Ibid.*)

Precipitation of Copper by Zinc. (*Ibid.*)

Persulphates of Rubidium, Calsium, etc. (*Ibid.*)

Electrolysis of Metallic Phosphates. (*Ibid.*)

Electrolysis of Metallis Phosphates. (*Ibid.*)

Observations upon Tungsten. (*Ibid.*)

The Electrolytic Oxidation of Toluene. (*Ibid.*)

The Separation of Molybdenum and Tungsten Trioxides. (*Ibid.*)

Richter's Organic Chemistry. Third American Edition, 2 Vols. (P. Blakiston's Son & Co., 1900.)

Richter's Inorganic Chemistry. Fifth American Edition. (P. Blakiston's Sons & Co., 1900.)

Smith & Keller's Chemical Experiments. Fourth Edition. (P. Blakiston's Son & Co., 1900.)

SOMMERVILLE, MAXWELL.

Grand Cameo of France. (Philadelphia.)

Hand-book of the Buddhist Temple.' (Philadelphia.)

The Græco-Roman Cameo Jupiter Ægiochus. (Published in 1898, but not previously reported.)

STEELE, J. DUTTON.

A Case of Typhoid Fever with Concurrent Malarial Infection. (*University Medical Magazine,* December, 1899.)

Endothelioma of Retroperitoneal Space. (*Proceedings of the Pathological Society of Philadelphia,* November, 1899.)

Melanotic Sarcoma of the Choroid, with Recurrence of the Growth in the Orbit. (*Proceedings of the Pathological Society of Philadelphia,* January, 1900.)

Some Cases of Dilatation of the Stomach. (*American Journal of the Medical Sciences,* February, 1900.)

Critical Summary of the Literature of Retroperitoneal Sarcoma. (*American Journal of the Medical Sciences,* March, 1900.)

Case of Contracted Kidneys, Thrombosis of the Coronary Arteries and Deformity of the Liver. (*Proceedings of the Pathological Society of Philadelphia,* April, 1900.)

The Pathology of Acquired Heart Disease in Children. (*Journal American Medical Association,* June 23, 1900.)

STENGEL, ALFRED.

Primary and Remote Effects of Athletics upon the Heart and Circulation. Read before Association of American Physicians, Washington, June, 1899. (*American Journal of Medical Sciences,* November, 1899. Lea Bros. & Co., Philadelphia.)

A New Operation for Gastroptosis. With Dr. H. D. Beyea. (*American Journal of Medical Sciences,* October, 1899. Lea Bros. & Co., Philadelphia.)

Aortic Regurgitation, with Remarks upon Flint's Murmur and Paroxysmal Sweating. (*Pennsylvania Medical Journal,* November, 1899.)

On the Pathology of the Erythrocyte. (*Contributions William Pepper Laboratory of Clinical Medicine,* 1900. W. B. Saunders & Co., Philadelphia.)

The Diagnosis of Chlorosis and Chloro-anemia. (*University Medical Magazine,* June, 1900.)

Diseases of Blood and of Ductless Glands and Metabolic Diseases. (Hare's *Progressive Medicine,* p. 103, June 1, 1900.)

STEVENS, ARTHUR A.

The Diagnosis of Mitral Stenosis. (*University Medical Magazine,* October, 1899.)

Materia Medica, Experimental Therapeutics, and Pharmacology. (*American Year Book of Medicine and Surgery,* 1900. Wm. B. Saunders & Co., Philadelphia.)

STEWART, COLIN C.

Zöllner's Anorthoscopic Illusion. (*American Journal of Psychology,* Vol. xi, No. 2, January, 1900, pp. 240-243.)

Mammalian Smooth Muscle. (*Proceedings of the American Physiological Society* in the *American Journal of Physiology,* Vol. iii, No. 8, March, 1900, pp. xxv-xxvi.)

The Reaction Time of Inhibition. With Allen Cleghorn. (*Ibid.,* pp. xxii-xxiii.)

Mammalian Smooth Muscle—The Cat's Bladder. (*American Journal of Physiology,* Vol. iv, No. 4. August, 1900, pp. 185-208.)

SWAN, JOHN M.

A Report of Five Cases of Diphtheria. (*Proceedings of the Philadelphia County Medical Society,* Vol. xix, Session 1898.)

A Manual of Human Anatomy. Arranged for First-year Students. (Philadelphia: Langmaid, 1900.)

Report of a Case of Chronic Valvular Endocarditis and Acute Nephritis following Chorea. (*Pediatrics,* Vol. ix, No. 9, 1900.)

Friedriech's Disease, with Report of a Fatal Case. With Dr. F. S. Pearce. (*Philadelphia Medical Journal,* May 26, 1900, Vol. v, No. 21.

William Harvey. (*Pennsylvania Medical Journal,* June, 1900.)

TRUMAN, JAMES.

Important Epochs in the Dentistry of the Nineteenth Century. (*International Dental Journal,* January, 1900.)

Dental Educational Methods. (*Ibid.,* February, 1900.)

Dental Literature of the Past and Present. (*Ibid.,* March, 1900.)

Is Dentistry a Charitable Profession? (*Ibid.,* May, 1900.)

Is Dentistry Drifting into Medicine? (*Ibid.,* July, 1900.)

Antisepsis in Dentistry. Chapter in *American Text Book of Operative Dentistry.* (Lea Brothers & Co., 1900, Philadelphia.)

TYSON, JAMES.

Articles on Diabetes and Nephritis in the *Cyclopaedia of Medicine and Surgery.* Compiled by Gould and Pyle. (P. Blakiston's Son & Co., January 15, 1900.)

The Cardio-vascular Changes in Bright's Disease. *Jacobi Festschrift,* May, 1900. Knickerbocker Press, New York.)

WALSH, JOSEPH.

A Sarcina Pathogenic for Guinea-pigs. (*Proceedings of Pathological Society of Philadelphia,* Vol. iii, No. 6, 1900.)

The Etiology of Pertussis: A Serum Treatment of Pertussis, with some Results. (*Contributions William Pepper Laboratory of Clinical Medicine,* Philadelphia, 1900, 450.)

WEBB, WALTER L.

Railroad Construction: Theory and Practice. A text-book for the use of students in colleges and technical schools. (Wiley & Sons, New York, 1900.)

Chapter on Progress in Civil Engineering in "Triumphs and Wonders of the Nineteenth Century." (A. J. Holman, Philadelphia, November, 1899.)

WESTCOTT, THOMPSON S.

Cerebro-spinal Symptoms in Influenza in Infancy. (*Archives of Pediatrics,* January, 1900, p. 8. New York: E. B. Treat & Co.)

WHITE, COURTLAND Y., JR.

Observations of Blood Changes Following Celiotomy. (*University Medical Magazine,* June, 1900.)

Tubercular Pneumonia in a Goat. With Dr. M. P. Ravenel. (*Transactions Philadelphia Pathological Society,* April, 1900.)

WHITE, J. WILLIAM.

Report of the Committee of the American Surgical Association on the Medico-Legal Relations of the X-Rays. (*American Journal of Medical Sciences,* July, 1900.)

Two Cases of Interscapulo-thoracic Amputation. (*University Medical Magazine,* May, 1900.)

WHITE, THOMAS RAEBURN.

Some Recent Criticism of Gelpcke *vs.* Dubuque. Sharswood Prize Essay for 1899 in the Law Department, University of Pennsylvania. (Privately printed in Philadelphia, December, 1899, by the University of Pennsylvania.)

WILLARD, DE FOREST.

Longitudinal Section of Femur from Pelvis, with Straightening for Ankylosis following Hip Disease. (*Annals of Surgery,* Vol. xxxi, p. 641; also *Transactions Academy of Surgery,* Vol. ii, p. 139, Philadelphia, 1900.)

Diagnosis and Treatment of Tuberculosis of the Bones and Joints. (*International Medical Magazine,* March, 1900, p. 175.)

Coxa Vara. (*University Medical Magazine,* June, 1900.)

Ankylosis Hip. (*Ibid.*)

Appendicitis. (*Cyclopaedia of Medicine and Surgery,* Gould & Pyle, 1900.)

Cystic Pelvic Tumor, Appearing below Poupart's Ligament, and Containing Numerous Peculiar Loose Bodies. With Dr. W. G. Spiller. (*Transactions Philadelphia Pathological Society,* 1900; also *Transactions American Surgical Association,* Vol. xviii, p. 355, and *University Medical Magazine,* October, 1900. Reprint.)

Exstrophy of the Bladder. (*Keating's Cyclopaedia,* Revised Edition.)

Sarcoma. Retroperitoneal Endothelioma. With Dr. J. D. Steele. (*Transactions Philadelphia College Physicians,* February 7, 1900; also *Transactions Philadelphia Pathological Society,* November, 1899, and *American Journal Medical Sciences,* March, 1900.)

Cholecystotomy; Suturing of Gall Bladder, without Drainage. (*Transactions Philadelphia Academy Surgery,* Vol. ii, p. 142, 1900.)

Pelvic Supporting Stool for Applying Gypsum Hip Dressings and Hip Spica Bandages. (*Annals of Surgery,* Vol. xxxi, p. 643; *Transactions Philadelphia Academy of Surgery,* Vol. ii, p. 141, 1900.)

WOOD, ALFRED C.

Acute Cholecystitis. (*University Medical Magazine,* November, 1899.)

WOOD, HORATIO C.

Therapeutics: its Principles and Practice. Eleventh edition. With Dr. Horatio C. Wood, Jr. (J. B. Lippincott Co., Philadelphia and London, 1900.)

The Plantar or Toe Reflex. Babinski's Symptom. (*University Medical Magazine,* April, 1900.)

President's Address at the Decennial Convention for the Revision of the United States Pharmacopœia. (*Philadelphia Medical Journal,* May, 1900.)

WOOD, HORATIO C., JR.

A Study of the Physiological Action of Phenacetin. With Dr. Harold B. Wood. (*University Medical Magazine,* July, 1900.)

A Comparative Study of Digitalis and its Derivatives. With Dr. John P. Arnold. (*American Journal of the Medical Sciences,* August, 1900.)

Therapeutics: its Principles and Practice. Eleventh edition. With Dr. H. C. Wood. (J. B. Lippincott Co., 1900.)

YOUNG, JAMES K.

Section on Hip Joint Disease. (*Cyclopaedia of Medicine and Surgery,* March, 1900.)

Treatment of Rhachitic Deformities. (*International Medical Magazine,* July, 1900.)

Treatment of Pott's Disease. (*Ibid.,* September, 1900.)

Treatment of Hare Lip. (*Ibid.,* November, 1900.)

YOUNG, JAMES T.

Liberty *versus* Efficiency. (*Yale Review*, November, 1899. Tuttle, Morehouse & Taylor, New Haven, Conn.)

The Administration of City Schools. (*Annals of the American Academy of Political and Social Science*, Philadelphia, March, 1900.)

Direct Legislation. (*Progress*, University Association, Chicago, Ill., June, 1900.)

TREASURER'S REPORT.

For the Year Ending August 31, 1900.

REPORT OF THE TREASURER OF THE UNIVERSITY OF PENNSYLVANIA.

THE PROPERTY OF THE UNIVERSITY

(part being held in trust and part in fee) is as follows, viz.:

GENERAL BALANCE SHEET.

Assets:

I. Real Estate*		$4,652,592 01	
II. Libraries, Museums, Apparatus, Furniture and Perpetual Insurance*		534,074 16	
III. Securities:*			
Railroad and Corporation Bonds	$1,672,654 78		
Stocks	230,068 80		
		1,902,723 58	
IV. Mortgages:			
On Property in Philadelphia and vicinity	$310,470 00		
On Property in Chicago and Western cities	328,000 00		
		638,470 00	
V. Ground Rents		63,820 62	
VI. Loans on Collateral		106,623 40	
			$7,898,303 77
VII. Cash Balances in various Trust Companies:			
To credit of Investment Funds:			
University Trusts		$15,363 15	
Hospital Trusts		3,765 03	
To credit of Building Funds		33,882 30	
To credit of Funds not for Investment:			
University		10,566 27	
Hospital		368 84	

*NOTE.—It is proper to state that Items I, II and III, above, have not been appraised within recent years. A reappraisal would add very largely to the estimated value, and such reappraisal will be made in the near future.

To credit of Income, University....	$2,416 47	
To credit of Administration, University	5,982 82	
		72,344 88
Total Assets		$7,970,648 65
Liabilities:		
Bonds Central Light and Heat Station	$80,000 00	
Mortgage, 3451 Woodland Avenue ..	5,500 00	
Mortgage, 3342 Walnut Street	5,000 00	
Obligations of General Fund	132,878 70	
Advances to Trusts by Administration	64,005 76	
Advances to Land purchases and Building Funds	220,859 74	
Loan to Veterinary Hospital	10,400 00	
		$518,644 20
		$7,452,004 45

REAL ESTATE.

IN PHILADELPHIA.

In Fifth Ward.

Premises No. 131 South Fifth Street	$17,000 00
Premises No. 224 Stampers Street	1,800 00

In Sixth Ward.

Premises No. 58 North Fourth Street	20,000 00
" " 60 " " "	30,000 00
" " 62 " " "	18,800 00
" " 64 " " "	14,600 00
" " 66 " " "	20,000 00
" " 66 " " " rear	6,000 00
" " 29 and 31 North Seventh Street	20,980 00

In Eleventh Ward.

Premises Nos. 429 and 431 North Second Street	7,239 76

In Twenty-seventh Ward.

Plot of ground between Thirty-fourth and Thirty-sixth Streets, Spruce Street and Woodland Avenue, with College Buildings and Medical Hall.................	643,799 27
Medical Laboratory	69,414 15
Plot of ground between Thirty-sixth, Spruce, Woodland Avenue, Woodland Cemetery and Guardian Street ...	250,000 00
Library Building	209,142 08
Veterinary Building	40,358 56
Biological Building	36,844 98

Plot of ground between Connecting Railroad and Thirty-fourth Street	$250,000 00
Plot of ground Thirty-third and Marston Street	12,764 46
Plot of ground between Thirty-sixth and Spruce Streets, Woodland Avenue and Police Station	20,000 00
John Harrison Laboratory of Chemistry	113,607 82
Henry Howard Houston Hall	134,544 57
Dormitory Building	344,268 67
New Dormitory Building, paid to August 31, 1900	134,983 27
Central Light and Heat Station	186,050 83
Department of Hygiene Building	50,000 00
Department of Hygiene Laboratory, paid to August 31, 1900	15,001 75
Dog Hospital, Veterinary Building	11,771 30
Bennett House, Nos. 3348 and 3350 Walnut Street	17,500 00
Premises No. 3451 Woodland Avenue	5,500 00
Department of Dentistry Building	129,851 08
Dining Hall	14,799 58
Law School Building, paid to August 31, 1900	373,488 52
Museum of Science and Art, paid to August 31, 1900	346,261 43
Premises Thirty-fourth and Locust Street, paid to August 31, 1900	112,510 90
Memorial Tower, paid to August 31, 1900	1,399 28
New Medical Laboratory, paid to August 31, 1900	100 00
Plot of ground Thirty-fourth to Thirty-sixth and Spruce to Pine Streets	125,000 00
Hospital Buildings	243,743 92
Gibson Wing	65,000 00
Nurses' Home	20,000 00
New Nurses' Home, paid to August 31, 1900	28,598 37
Mortuary Building	11,000 00
Maternity Building	29,000 00
Agnew Memorial Pavilion	166,645 57
Pepper Clinical Laboratory	25,000 00
James Dundas Clinic	6,098 67
Premises Nos. 3340, 3342, 3344, 3346 Walnut Street	30,000 00
In Twenty-ninth Ward.	
Premises No. 3011 Girard Avenue	5,300 00
In Thirty-ninth Ward.	
Twenty-three and one-half acres on Delaware River	100,000 00
Two and eighty-seven one hundredths acres on Moore Street	20,000 00
In Delaware County, Pennsylvania.	
Reese Wall Flower Estate, Farm	30,000 00
" " " " Observatory Buildings	12,796 84
" " " " Professor's House and Library	11,808 25

IN ELK AND MCKEAN COUNTIES, PENNSYLVANIA.	
About seven hundred acres	$5,054 71
IN LIGONIER, WESTMORELAND COUNTY, PENNSYLVANIA.	
Sundry lots of ground	7,000 00
SEA ISLE CITY, N. J.	
Marine Biological Laboratory and grounds	4,000 00
IN MANCHESTER, MASS.	
About sixty-four acres	26,163 42
Total Real Estate	$4,652,592 01

LIBRARIES, MUSEUMS, APPARATUS, FURNITURE AND PERPETUAL INSURANCE.

University Libraries	$123,500 00
College Apparatus	50,000 00
Museum of Archæology and Paleontology	97,000 00
Wistar and Horner Museum	40,000 00
Museum of Biology	5,000 00
Dr. George B. Wood Museum	25,000 00
Laboratory of Hygiene	19,874 28
J. H. Towne Collection of Fossils	4,975 00
Flower Observatory	24,098 78
Furniture:	
Medical Hall	10,302 90
Medical Hall, Ventilating Apparatus	24,086 90
Veterinary Building	3,867 38
Hospital	15,511 43
Department of Biology	883 22
Library Building	10,511 62
College Building	7,500 00
Houston Hall	22,891 21
Department of Dentistry	16,748 01
Dormitory	6,747 91
Dining Hall	522 42
Perpetual Insurance:	
On College Buildings	10,680 00
" Veterinary Building and Dog Hospital	900 00
" Hospital Building	4,947 50
" Gibson Wing Building	900 00
" Biological Building	300 00
" Central Light and Heat Station	1,000 00

On Department of Hygiene Building	$875 00
" Chemical Laboratory Building	1,200 00
" Department of Dentistry Building	2,565 00
" New Nurses' Home Building	900 00
" Flower Observatory and Professor's House	200 00
" Trust Properties in City	585 60
	$534,074 16

SECURITIES.

RAILROAD AND CORPORATION BONDS.

5,000	Chicago Gas Light & Coke Co. 1st Mtg. 5's, 1937	$5,000 00
61,000	Reading Co. and P. & R. Coal and Iron Gen. Mtg. 4's, 1997	47,817 50
3,000	Philadelphia & Reading R. R. Cons. 7's, 1911	3,000 00
25,000	" " " " 4's, 1937	25,000 00
12,000	Elmira & Williamsport R. R. Co. Income Bonds, 5's, 2862	12,000 00
2,000	Pennsylvania R. R. Co. Cons. Mtg. 6's, 1905	1,924 00
1,000	Susquehanna Coal Co., Guar. by Pennsylvania R. R. Co. 6's, 1911	1,000 00
25,000	Toledo, O., Electric Street Railway Co. 1st Mtg. 5's, 1912	25,000 00
5,000	Metropolitan St. Ry, Kansas City, Mo., Gen. Mtg. 5's, 1909	4,894 17
30,000	Evansville & Terre Haute R. R. Co. 1st Mtg. 5's, 1942	30,600 00
17,000	Bethlehem Iron Co. 1st Mtg. 5's, 1907	17,000 00
50,000	International Nav. Co. 1st Mtg. 5's, 1929	50,000 00
13,000	Norfolk & Western R. R. Co. 1st Mtg. 6's, 1932	13,000 00
58,000	Atlanta Gas Light Co. 1st Mtg. 5's, 1947	56,750 00
2,000	Catawissa R. R. Co. Extended 6's, 1902	2,000 00
101,000	San Antonio & Aransas Pass. R. R. 1st Mtg. 4's, 1943	60,000 00
2,500	County of St. Louis, Minn., 6's, 1907	2,500 00
2,500	Wilmington, Columbia & Augusta R. R. 1st Mtg. 6's, 1910	2,500 00
1,350	Lehigh Coal & Navigation Co. Cons. Loan 4½'s	1,471 50
5,000	Columbus & Indianapolis Central R. R. 2d Mtg. 7's, 1904	5,000 00
13,000	Wabash R. R. Co. 2d Mtg. 5's, 1939	13,000 00
2,000	Saltzburg Coal Co. Gen. Mtg. 6's, 1914	2,000 00
3,000	Pittsburg, Shenango & Lake Erie R. R. 1st Mtg. 5's, 1940	3,000 00

5,000	Fairmount Coal & Coke Co., 1st Mtg. 4's, 1919....	$5,000 00
150,000	Philadelphia & Reading Coal & Iron Co. 5-10-year 6's, 1904	152,500 00
22,000	City of Galveston, Tex., 40-year Ltd. 5's, 1934....	22,000 00
25,000	Lehigh Valley R. R. Co. Mtg. and Collat. Trust 5's, 1997	25,000 00
53,000	Erie R. R. Co. prior lien 4's, 1996	48,082 50
25,000	Scranton Traction Co. 1st Mtg. 6's, 1932	25,000 00
22,000	Ohio River R. R. Co. 1st Mtg. 5's, 1936	22,660 00
98,000	Equitable Ill. Gas L. Co. 1st Mtg. 5's, 1928.......	99,440 00
12,000	Choctaw, Oklahoma & Gulf R. R. Co. Gen. Mtg. 5's, 1919	12,050 00
10,000	Raleigh & Gaston R. R. 1st Mtg. 5's, 1947	10,332 50
10,000	Syracuse Rapid Transit Co. 1st Mtg. 5's, 1946....	10,000 00
11,000	Western Gas Co. 1st Mtg. 5's, 1933	10,987 50
3,500	Union Pacific R. R. 1st Mtg. Land Grant 4's, 1947	3,500 00
40,000	Pittsburg Consol. Gas. Co. 1st Mtg. 5's, 1948	41,000 00
3,300	Bloomsburg & Sullivan R. R. 1st Mtg. 5's, 1928...	3,300 00
1,600	" " " 2d Mtg. 5's, 1928 ...	1,600 00
66.66	" " " Scrip	66 66
22,000	Kansas City, Mo., Gas. Co. 1st Mtg. 5's, 1922	21,882 50
500	Philadelphia Country Club 1st Mtg. 5's, 1940	500 00
2,000	Buffalo & Susquehanna R. R. 1st Mtg. 5's, 1913 ..	1,990 00
21,000	Second Avenue (Pittsburg) Traction Co. 1st Mtg. 5's, 1934	21,000 00
30.000	Chicago & Erie R. R. 1st Mtg. 5's, 1982	31,975 00
2,000	Brooklyn, N. Y., Union Gas. Co. 1st Mtg. 5's, 1945	2,000 00
30,000	Lehigh Valley R. R. Car Trust 5's, 1900-1904....	30,000 00
30,000	Central Light and Heat Station 5 4-10's	30,000 00
132,878.70	Obligation of the University	132,878 70
34,000	Lehigh Valley Coal Co. 1st Mtg. 5's, 1933	33,850 00
25,000	St. Louis Merchants' Bridge Co. 1st Mtg. 6's, 1929	25,000 00
10,000	Sunbury, Hazleton & Wilkesbarre R. R. 1st Mtg. 5's, 1928	9,380 00
400	Sunbury, Hazleton & Wilkesbarre R. R. 2d Mtg. 6's, 1938	393 50
25,000	St. Louis Terminal Cupples Station and Property 1st Mtg. 4½'s, 1917	24,468 75
25,000	Consumers' Light, Heat & Power Co. 1st Mtg. 5's, 1938	25,000 00
1,000	City of Pittsburg Refunding 5's, 1910	1,130 00
1,000	Central R. R. of N. J. Gen. Mtg. 5's, 1987	1,120 00
6,000	Philadelphia & Erie R. R. Gen. Mtg. 5's, 1920	7,080 00
25,000	Citizens' Ry. Indianapolis 1st Con. Mtg. 5's, 1933..	24,000 00
25,000	Philadelphia & West Chester Traction 1st Mtg. 5's, 1918	25,000 00
10,400	Veterinary Hospital Bonds	10,400 00

27,000	Wilmington & Chester Traction Co. Collat. Trust 5's, 1918	$27,000 00
21,000	Williamsport Gas. Co. 1st Mtg. 5's, 1939	21,000 00
25,000	North Hudson Heat, Light & Power Co. Mtg. 5's, 1938	25,000 00
25,300	Electric & Peoples' Traction Trust Certificate 4's, 1945	25,028 75
20,000	Baltimore & Ohio R. R. 1st Mtg. 4's, 1948........	19,925 00
10,000	Baltimore & Ohio R. R. prior lien 3½'s, 1925	9,412 50
30,000	Hudson County Gas Co. 1st Mtg. 5's, 1949	30,000 00
10,000	Northern Pacific R. R. prior lien 4's, 1997	10,000 00
60,000	Jersey City, Hoboken & Paterson Street Ry. 1st Mtg. 4's, 1949	50,000 00
25,000	United Railways Co. Gold Trust 4's, 1949	21,250 00
25,000	Norfolk & Western R. R. Co. 1st Cons. Mtg. 4's, 1996	24,437 50
35,000	Western N. Y. & Penna. R. R. Gen. Mtg. 3-4's, 1943	32,462 50
20,000	St. Louis, Iron Mountain & Southern R. R. Land Grant 5's, 1931	20,000 00
5,000	Phila. Electric Gold Trust 5's, 1948	4,993 75
6,000	Lehigh Valley R. R. Car Trust 4½'s, 1910	6,120 00
	Total Railroad and Corporation Bonds	$1,672,654 78

STOCKS.

16	shares	Lehigh Valley R. R. Co. pfd................	$800 00
120	"	Wm. Cramp & Sons Ship & Engine Bldg. Co..	6,000 00
61	"	Norfolk & Western R. R. Co. Com..........	5,550 00
507	"	Lehigh Valley R. R. Co. Com...............	15,690 00
5	"	American Dredging Co....................	500 00
10	"	Lehigh Coal and Navigation Co.............	500 00
500	"	Delaware & Bound Brook R. R...............	75,000 00
267	"	Pennsylvania R. R. Co....................	14,027 05
27	"	Brooklyn Union Gas Co. pfd................	2,160 00
238	"	Phila. & Reading R. R. 2d pfd...............	2,856 00
578	"	Phila. & Reading R. R. Com.................	5,780 00
67	"	Safe Deposit & Trust Co., Pittsburg........	4,020 00
25	"	Union Pacific R. R. pfd.....................	1,500 00
10	"	American Snuff Co........................	1,000 00
80	"	Fidelity Insurance Trust & Safe Deposit Co..	35,600 00
22	"	Provident Life & Trust Co..................	9,482 00
49	"	Pennsylvania Salt Mfg. Co..................	5,390 00
500	"	United Traction Co., Pittsburg Pref........	25,000 00

200 Shares Phila. Traction Co.	$19,183 75
3 " Insurance Co. of North America	30 00
Total Stocks	$230,068 80

MORTGAGES

ON PROPERTY IN PHILADELPHIA AND VICINITY.

No. 707 Lydia Street	$700 00
Nos. 752 and 754 South Eighth Street	4,000 00
Wallace and Nineteenth Street	6,000 00
Lancaster Avenue and Rockland Street	2,750 00
Cheltenham Township, Montgomery Co., Pa.	10,000 00
Cheltenham Township, Montgomery Co., Pa.	3,000 00
Gray's Ferry Road and Thirtieth Street	2,000 00
Jenkintown, Pa.	4,320 00
Auburn and Tenth Street	9,000 00
Washington Avenue and Twenty-first Street	35,000 00
No. 4044 Powelton Avenue	3,000 00
Nos. 2262, 2318 and 2320 North Front Street	8,000 00
Northeast corner Woodland Ave. and Thirty-fourth Street	2,500 00
No. 1343 Mount Vernon Street	2,700 00
No. 2977 Frankford Avenue	4,000 00
No. 2987 Frankford Avenue	4,000 00
Camden, N. J.	10,000 00
Thirty-fifth Ward, Philadelphia	80,000 00
No. 1730 Pine Street	5,000 00
Glenloch, Chester Co., Pa.	38,000 00
Woodland Avenue and Sixty-second Street	50,000 00
City Line and Washington Lane	14,000 00
No. 549 North Second Street	3,000 00
No. 3342 Walnut Street	5,000 00
Manchester, Mass.	1,000 00
Ligonier, Pa.	3,500 00
Total	$310,470 00

COUPON MORTGAGES.

In Chicago, Illinois	$20,000 00
" " "	8,500 00
" " "	30,000 00
" " "	8,000 00
" " "	2,500 00
" " "	10,000 00
" " "	25,000 00
" " "	25,000 00

In Chicago, Illinois	$10,000 00
" " "	6,000 00
" " "	20,000 00
" " "	10,000 00
" " "	6,000 00
" " "	10,000 00
" " "	2,500 00
" " "	2,500 00
" " "	35,000 00
" " "	15,000 00
" Kansas City, Missouri	20,000 00
" " " "	20,000 00
" Olympia, Wash.	6,000 00
" Lincoln, Neb.	30,000 00
" Atlanta, Ga.	6,000 00
Total	$328,000 00

GROUND RENTS IN PHILADELPHIA, ETC.

No. 29 North Sixth Street	$2,500 00
No. 533 North Second Street	426 67
No. 535 North Second Street	426 67
No. 537 North Second Street	426 67
Green and New Market Streets	111 11
Moore and Otsego Streets	3,000 00
Oakford, Manton and Twenty-second streets, nine at $800 each	7,200 00
Stiles Street and Nineteenth Street	900 00
Gray's Ferry Road	42,519 00
No. 429 North Second Street	2,400 00
No. 110 Spruce Street	1,866 67
Nos. 13 and 15 Parkham Street	400 00
No. 618 South Seventh Street	444 50
No. 128 North Water Street	1,133 33
Lebanon, Pa	66 00
Total Ground Rents	$63,820 62

Loan on Collateral	$50,000 00
" " "	1,500 00
" " "	20,000 00
" " "	10,123 40
" " "	25,000 00
Total Loan	$106,623 40

THE VARIOUS TRUST FUNDS HELD BY THE UNIVERSITY ARE INVESTED IN REAL ESTATE AND OTHER SECURITIES ENUMERATED IN THIS REPORT.

The following is a list of the various Trusts:

Endowment Funds—University.	Amount Invested.	Amount Uninvested.
General Fund	$2,027,079 69	$119 08
J. Herman Raht Memorial Fund	250 00	
Phi Kappa Sigma Fund	400 00	
Keble Fund	12,868 41	
Elliott Cresson Fund	6,701 65	
Tobias Wagner Fund	5,300 00	
Bloomfield Moore Fund	15,000 00	
Thomas A. Scott Fund	50,000 00	
John H. Towne Fund	426,021 88	
John H. Towne Fossils Fund	5,000 60	
John Welsh Fund	49,822 24	2,000 00
A. Whitney Fund	44,542 75	5,401 25
Scholarship Fund	6,084 85	52 22
Benjamin Franklin Scholarship Fund	10,026 12	
Samuel V. Merrick Scholarship Fund	3,508 10	
Matthew Baird Scholarship Fund	5,013 06	
A. Seybert Fund	53,045 74	
Chair of Christian Ethics Fund	3,015 36	
Alumni Hall Fund	5,900 44	42 67
H. Tyndale Scholarship Fund	12,194 78	
S. W. Roberts Fund	1,114 83	
Wharton School Fund	157,754 00	
Assyrian Fellowship Fund	189 55	
Pepper Professorship of Hygiene Fund	59,050 00	350 00
J. Rhea Barton Fund	49,920 48	
J. Gillingham Fell Fund	1,100 00	
Auxiliary Faculty Medicine Fund	31,269 00	71 16
Joseph Leidy Chair Fund	1,837 16	
Botanic Garden Fund	4,302 99	
Department of Biology Fund	47,038 44	
I. V. Williamson Fund	47,530 00	
George S. Pepper Residue Legacy Fund	34,031 37	
Francis S. Pepper Scholarship Fund	6,262 50	887 50
Chair of Pedagogy Fund	250 00	
Henry Reed Prize Fund	743 12	
Latin Prize Fund	1,025 80	

Endowment Funds—University.	Amount Invested.	Amount Uninvested.
Oratory Prize Fund	$360 80	
English Composition Prize Fund	200 00	
J. Warner Yardley Prize Fund	500 00	
Class of 1880 Prize Fund	1,000 00	
Allen Memorial Prize Fund	886 25	$113 75
Law School Prize Fund	1,000 00	
Krauth Library Fund	1,604 64	
Agricultural Library Fund	500 00	
Dr. Pepper Library Fund	7,708 19	
Tobias Wagner Library Fund	10,087 08	
Lippincott Library Fund	7,000 00	
Norris Library Fund	5,233 69	
Reese Wall Flower Fund	109,213 87	
John Harrison Lab. of Chemistry Fund...	114,829 82	
Veterinary Fund	75,375 94	
J. M. Bennett Endowment Fund	17,500 00	
Howard Houston Hall Fund	157,435 78	
Howard Houston Hall Endowment Fund	500 00	
Dormitory Fund	355,016 58	
Dining Hall Fund	14,799 58	
Dental Hall Fund	149,164 09	
Wistar and Horner Museum Fund	40,110 11	
Department of Law Fund	7,735 08	
John F. Fraser Fund	6,332 64	
Department of Hygiene Fund	189,388 44	
Thomas A. Scott Fellowship Fund	10,000 00	
J. M. Bennett End. Grad. Dept. for Women	45,737 50	
A. Sidney Biddle Fund	20,245 13	4 87
Simon Muhr Scholarship Fund	9,525 60	
George L. Harrison Foundation Fund ...	558,216 25	1,572 11
John Stewardson Memorial Fund	17,771 56	
Sinking Fund	1,261 05	3,691 04
John White Field and Francis Peters Scholarship Fund	10,000 00	
Pepper Hall Endowment Fund	49,096 25	903 75
James Logan Scholarship Fund	5,000 00	
Robert Morris Scholarship Fund	5,000 00	
Garrett Scholarship Fund	9,861 25	138 75
Louise Harrison Scholarship Fund	5,122 50	
University Fund for Women	100 00	
Professors' Retiring Fund	1,985 00	15 00
Total University Endowments	$5,207,599 58	$15,363 15

Endowment Fund—Hospital.	Amount Invested.	Amount Uninvested.
General Fund	$834,075 87	$2,265 76
Hahn Ward Fund	54,650 00	39 75
Alumni Ward Fund	10,017 19	8 06
Mrs. G. W. Norris Fund	10,000 00	
Bement Fund	5,000 00	
Ward Chronic Diseases Fund	196,967 50	129 06
Dr. Pepper Clinical Laboratory Fund	25,030 00	
I. V. Williamson Hospital Fund	47,266 81	33 19
Children's Ward Fund	25 00	7 07
Henry Seybert Fund	60,317 31	45 45
George S. Pepper Fund	72,729 19	103 14
Recto-Genito-Urinary Fund	4,060 00	40 39
Anna M. Powers Bed Fund	6,000 00	
Mary W. Eskens Bed Fund	5,000 00	
Dr. D. Hayes Agnew Hospital Fund	171,645 57	
Dr. D. Hayes Agnew Bequest for Care of Children Fund	49,310 00	690 00
Dr. D. Hayes Agnew Bequest for Maternity Hospital	1,000 00	
Andrew C. Butler Fund	5,415 00	17 67
Ligonier Free Bed Fund	37,739 31	195 62
Children's Orthopædic Ward Fund	14,140 00	
Nurses' Home Building Fund	450 00	
Ladies' Fund for Maternity Hospital	344 31	13 13
A. Sidney Roberts Apparatus Fund	6,249 06	15 48
John White Field Bed Fund	4,985 00	15 00
Abby White Peters Bed Fund	4,986 00	14 00
Maternity Fund	9,965 00	119 00
Edward Rhoads Memorial Fund	3,398 50	7 01
Alfred Gustavus Baker Bed Fund	4,993 75	6 25
Abraham and Mary Baker Bed Fund	5,100 00	
Total Hospital Endowment	$1,650,860 37	$3,765 03

New Building Funds.	Amount Invested.	Amount Uninvested.
Laboratory of Hygiene	$15,001 75	
Free Museum of Science and Art	346,261 43	
New Medical Laboratory	100 00	$12,893 06
New Dormitories	184,883 27	
Law School	373,588 52	
Memorial Tower	1,399 28	20,989 24
Premises Thirty-fourth and Locust Streets	112,510 90	
James Dundas Clinic	6,098 67	
Total	$1,039,843 82	$33,882 30

SUMMARY.

Investments—University Trusts	$5,207,599 58
Hospital Trusts	1,650,860 37
New Buildings	1,039,843 82
Total Investments	$7,898,303 77

Income Account.	Amount received including Balance August 31, 1899.	Amount Expended.	Balance.
General Fund	$1,822 68	$1,812 17	$10 51
J. Herman Raht Fund	10 00	10 00	
Phi Kappa Sigma Fund	16 00	16 00	
Keble Fund	778 86	778 86	
Elliott Cresson Fund	28 08	28 08	
Tobias Wagner Fund	212 00	212 00	
Bloomfield Moore Fund	971 04	971 04	
Thomas A. Scott Fund	2,471 36	2,471 36	
J. H. Towne Fund	13,741 82	13,741 82	
J. H. Towne Collection Fund.	1 28	1 28	
John Welsh Fund	2,519 91	2,519 91	
A. Whitney Fund	2,611 21	2,611 21	
Scholarship Fund	165 62	165 62	
Benjamin Franklin Scholarship Fund	451 30	451 30	
Samuel V. Merrick Scholarship Fund	175 40	175 40	
Matthew Baird Scholarship Fund	300 66	300 66	
A. Seybert Fund	2,714 78	2,714 78	
Chair Christian Ethics Fund	150 76	150 76	
Alumni Hall Fund	295 33	295 33	
Tyndale Scholarship Fund	616 41	616 41	
S. W. Roberts Fund	48 14	48 14	
Wharton School Fund	8,686 00	8,686 00	
Assyrian Fellowship Fund	9 48	9 48	
Pepper Professorship of Hygiene Fund	3,476 38	3,476 38	
J. Rhea Barton Fund	2,198 86	2,198 86	
J. Gillingham Fell Fund	2 00	2 00	
Aux. Faculty Medicine Fund.	1,643 59	1,643 59	
Leidy Chair Fund	64 96	64 96	
Botanic Garden Fund	92 14	92 14	
Department Biology Fund	52	52	
I. V. Williamson Fund	2,495 43	2,495 43	
George S. Pepper Residuary Legatee Fund	1,518 36	1,518 36	

Income Account.	Amount received including Balance August 31, 1899.	Amount Expended.	Balance.
Frances S. Pepper Scholarship Fund	$566 87	$566 87	
Chair of Pedagogy Fund	12 82	12 82	
Henry Reed Prize Fund	47 83	25 00	$22 83
Latin Prize Fund	58 11	43 00	15 11
Oratory Prize Fund	61 53	55 00	6 53
English Composition Prize Fund	21 32		21 32
J. Warner Yardley Prize Fund	56 21	25 00	31 21
Class 1880 Prize Fund	50 17		50 17
Allen Memorial Prize Fund	109 83	80 00	29 83
Law School Prize Fund	20 14	20 00	14
Rev. C. P. Krauth Library Fund	245 03	127 03	118 00
Agricultural Library Fund	62 50	49 72	12 78
Dr. Pepper Library Fund	745 67	583 59	162 08
Tobias Wagner Library Fund	924 90	906 01	18 89
Lippincott Library Fund	582 50	329 00	253 50
Norris Library Fund	678 94	594 87	84 07
R. W. Flower Fund	1,460 66	1,259 66	201 00
Veterinary Fund	843 15	843 15	
Houston Hall Endow. Fund	25 62	25 62	
Dormitory Fund	200 00	200 00	
Wistar and Horner Mus. Fund	5 50	5 50	
Department of Law Fund	309 76	309 76	
John F. Fraser Fund	350 02	350 02	
Department Hygiene Fund	6,218 76	6,218 76	
Thomas A. Scott Fellowship Fund	450 00	450 00	
J. M. Bennett Endowment Graduate Dept. Fund	2,155 95	2,155 95	
A. S. Biddle Fund	952 00	952 00	
Simon Muhr Fund	361 28	361 28	
George L. Harrison Foundation Fund	30,391 53	30,391 53	
Department Archæology and Paleontology Fund	250 00	250 00	
John Stewardson Memorial Fund	1,997 30	772 90	1,224 40
Sinking Fund	17 50	17 50	
New Dormitories Building Fund	1,200 00	1,200 00	
Law Dept. Building Fund	1,000 00	1,000 00	

Income Account.	Amount received including Balance August 31, 1899.	Amount Expended.	Balance.
Field Scholarship Fund	$432 57	$432 57	
Pepper Hall Endowment Fund	1,079 00	1,044 90	$34 10
Jas. Logan Scholarship Fund	125 00	125 00	
Robt. Morris Scholarship Fund	125 00	125 00	
Garrett Scholarship Fund ...	230 00	230 00	
Louise Harrison Scholarship Fund	120 00		120 00
	$104,835 33	$102,418 86	$2,416 47

Less Balances August 31, 1899	$3,039 22	
Less Interest added to Income	49 22	
Less Contributions added to Income	86 18	
		3,174 62
Net Income for the Year,		$101,660 71

Funds not for Investment.	Amount received, including Balance August 31, 1899.	Amount Expended.	Balance
A. Seybert Library Fund	$2 50	$2 50	
Mediæval Philosophy Library Fund	370 95	86 58	$284 37
Warren Library Fund	1,000 00	524 15	475 85
Comegys Library Fund	40 14	37 16	2 98
Special No. 3 Library Fund..	394 62	255 92	138 70
Architecture Library Fund...	7 75	7 75	
Musical Library Fund	257 84		257 84
Kendall Library Fund	25 68	1 60	24 08
Pathology Library Fund	25 00		25 00
Law Library Fund	5,640 46	4,821 33	819 13
Blue Book Library Fund	1,842 60	1,313 45	529 15
Dramatic Library Fund	17 00	17 00	
Cur. Expense Library Fund..	578 36	578 36	
John C. Ropes Prize Fund...	4 95		4 95
Comegys Prize Fund	50 00	50 00	
Delano Prize Fund	50 00	50 00	
Mrs. Harrison's Fund for Casts	26 27		26 27
Architecture Lecture Fund ..	236 75	43 58	193 17

Funds not for Investment.	Amount received, including Balance August 31, 1899.	Amount Expended.	Balance.
Architecture Equipment Fund	$4,130 94	$1,552 16	$2,578 78
Architecture, General Fund for Casts	71 52		71 52
Architecture, Peabody Fund..	103 17		103 17
Civil Engineering Equipment Fund	514 33	514 33	
E. B. Coxe, Jr., Fund	3,608 29	3,608 29	
University Publication Fund.	3,866 73	3,608 11	258 62
'87's House Dormitory Fund.	879 00	564 00	315 00
Improvement Mechanical and Elec. Engineering Fund..	1,567 80	1,460 28	107 52
Lecture Course State of Pennsylvania Fund	313 55	32 50	281 05
Alumnæ Fellowship for Women Fund	850 00	250 00	600 00
Debate Team 1899 Fund	7 24	7 24	
Fellowship in English Fund..	970 00	470 00	500 00
Vivarium Fund	5,283 44	5,283 44	
Orchid Collection Fund	225 50	225 50	
Gymnasium Fund	6,500 00	6,500 00	
Needy Students' Fund	4,117 08	4,117 08	
Cur. Expense University Fund	70,105 00	70,105 00	
Samuel Jackson Mem. Fund.	2,880 00	2,880 00	
"Several Friends of the University"	40,000 00	40,000 00	
Locust Street Improvement Fund	990 00	945 62	44 38
Opening Ceremonies Law School Fund	3,014 59	2,569 39	445 20
Scholarship Depart. Biology Fund	240 00	240 00	
Dr. Nathaniel Chapman Mem. Fund	2,850 00	2,850 00	
Semitic Department Fund	100 00	100 00	
Mechanical Eng. Dept. Adver. Fund	300 00	264 85	35 15
Lippincott Veterinary Fund..	2,500 00	2,500 00	
Physical Laboratory Fund	25,000 00	25,000 00	
American School at Rome Fund	450 00	450 00	
Y. M. C. A. Univ. of Penna. Fund	500 00		500 00
N. Y. Alumni House Fund..	4,500 00	4,500 00	

Funds not for Investment.	Amount received including Balance August 31, 1899.	Amount Expended.	Balance
Electro-Chem. App. Fund...	$1,850 00		$1,850 00
Suspense Fund	372 10	$277 71	94 39
			$10,566 27

UNIVERSITY TRUSTS.

GENERAL BALANCE SHEET.

RECEIPTS AND EXPENDITURES FROM AUGUST 31, 1899, TO AUGUST 31, 1900, INCLUSIVE.

Balance, August 31, 1899			$145,867 41
Receipts:			
Income from Invested Funds		$101,660 71	
Donations for Endowment...	$322,466 00		
Donations not for Endowment	145,209 35		
		467,675 35	
Securities Sold and Redeemed...........		143,362 09	
Premium received		50 00	
Advances by Administration		54,968 69	
From Departments Account Sinking Fund and Bond Interest		11,601 00	
Interest on Deposits		2,198 62	
Transfer to Trusts from Administration..		5,532 00	
Addition to Capital Account		6,956 89	
Miscellaneous		534 35	
			794,539 70
Total Receipts			$940,407 11

Expenditures:	
Investments made	$264,924 93
Accrued Interest and Premiums paid.....	6,121 89
Buildings and Grounds	419,586 06
Taxes, Water Rent, Insurance, Legal Expenses	5,310 68
Funds not for Investment transferred to Administration	76,380 79
Income transferred to Administration	68,419 82
Advances to Trusts repaid	42,663 44
Paid on account Sundry Trusts as per Schedule	14,553 40

Interest	$4,072 50	
Bonds paid	10,000 00	
Suspense Account	277 71	
		$912,311 22
Balance, August 31, 1900		$28,095 89

BUILDINGS AND GROUNDS.

Premises Thirty-fourth and Locust Streets	$98,985 90	
Old Dormitory	590 12	
Laboratory of Hygiene	5,151 75	
Law School Building	170,947 63	
New Dormitory	130,207 65	
Free Museum of Science and Art	6,920 29	
Memorial Tower	1,399 28	
New Medical Laboratory	100 00	
Vivarium	5,283 44	
		$419,586 06

PAYMENTS ON ACCOUNT SUNDRY TRUSTS.

John Stewardson Memorial Fund	$772 90	
Blue Book Library Fund	1,313 45	
Opening Ceremonies Law School Fund	2,569 39	
Architecture Lecture Fund	43 58	
Mechanical Engineering Dept. Adver. Fund	264 85	
Architecture Equipment Fund	840 25	
Locust Street Improvement Fund	945 62	
Lecture Course State of Pennsylvania Fund	32 50	
Orchid Collection Fund	225 00	
Needy Students Fund	1,705 00	
University Publication Fund	3,608 11	
American School at Rome Fund	450 00	
Reese Wall Flower Fund	319 47	
Pepper Hall Endowment Fund	3 00	
Improvement Mechanical and Electrical Engineering Department Fund	1,460 28	
		$14,553 40

ADMINISTRATION ACCOUNT.

RECEIPTS AND EXPENDITURES FROM AUGUST 31, 1899, TO AUGUST 31, 1900, INCLUSIVE.

Balance, August 31, 1899			$35,926 90
Receipts:			
From Tuition Fees		$323,120 60	
" Rents and Charges of Dormitories		28,650 15	
" Graduation Fees		2,830 00	
" Income from Invested Funds		68,419 82	
" Funds not for Investment		76,380 79	
" Departments:			
Veterinary Hospital	$8,620 00		
Central Light and Heat Station	27,254 88		
Dept. of Repairs	13,942 95		
		49,817 83	
" Advances to Trusts repaid		72,768 91	
" Sinking Fund		10,000 00	
" Interest		941 46	
" Miscellaneous		136 40	
			633,065 96
			$668,992 86
Expenditures:			
Salaries, Educational		$295,579 60	
Salaries, Administration		68,636 14	
Current Expense Account		171,777 44	
Advances to Trusts		85,074 16	
Bills Payable		30,000 00	
Transfer to Trusts		5,532 00	
Interest paid		6,410 70	
			663,010 04
Balance, August 31, 1900			$5,982 82

ANALYSIS OF ADMINISTRATION ACCOUNT.

			Surplus.	Deficit.
College:				
Tuition Fees	$93,227 15			
Income	39,566 50			
		$132,793 65		
Salaries, Education ..	$151,993 46			
Salaries, Admin......	8,924 55			
Cur. Expense Acc't..	42,718 53			
		203,636 54		
				$70,846 89
Department of Medicine:				
Tuition Fees	$124,900 00			
Income	2,200 86			
		$127,100 86		
Salaries, Education ..	$67,871 42			
Salaries, Admin......	7,342 98			
Cur. Expense Acc't..	46,977 07			
		122,191 47		
			$4,909 39	
Department of Law:				
Tuition Fees	$46,100 00			
Income	952 00			
		$47,052 00		
Salaries, Education ..	$28,149 36			
Salaries, Admin.....	2,478 72			
Tran. to Law Library	5,532 00			
Cur. Expense Acc't..	14,512 92			
		50,673 00		
				3,621 00
Department of Dentistry:				
Tuition Fees		$58,472 60		
Salaries, Education..	$23,390 65			
Salaries, Admin......	3,008 46			
Cur. Expense Acc't..	34,474 08			
		60,873 19		
				2,400 59
Department of Hygiene:				
Tuition Fees	$110 00			
Income	5,742 73			
		$6,852 73		

			Surplus.	Deficit.
Salaries, Education ..	$4,800 00			
Salaries, Admin......	480 00			
Cur. Expense Acc't..	1,963 90			
		$7,243 90		
				$391 17
G. L. Harrison Foundation:				
Income		$13,130 63		
Salaries, Education .		13,130 63		
Department Veterinary Medicine:				
Tuition Fees		$3,277 25		
Salaries, Education..	$3,900 00			
Salaries, Admin.....	1,567 62			
Cur. Expense Acc't..	1,490 36			
		6,957 98		
				3,680 73
Veterinary Hospital:				
Receipts	$8,620 00			
Income	573 15			
		$9,193 15		
Salaries, Admin.....	$4,610 31			
Cur. Expense Acc't..	4,590 21			
		9,200 52		
				7 37
Department of Repairs:				
Receipts		$13,942 95		
Salaries Admin......	$3,054 43			
Cur. Expense Acc't..	7,603 94			
		10,658 37		
			3,284 58	
Dormitories:				
Receipts		$27,736 66		
Salaries Admin......	$5,085 60			
Cur. Expense Acc't..	8,117 72			
		13,203 32		
			14,533 34	
Bennett Hall:				
Receipts	$913 49			
Income	1,876 92			
		$2,790 41		
Salaries Admin......	$926 00			
Cur. Expense Acc't..	2,718 93			
		3,644 93		
				854 52

			Surplus.	Deficit.
Central Light and Heat Station:				
Receipts		$27,254 88		
Salaries Admin......	$7,696 87			
Cur. Expense Acc't..	22,297 61			
		29,994 48		$2,739 60
			$22,727 31	$84,537 87

SUMMARY.

Deficits:		
College	$70,842 89	
Department of Law......................	3,621 00	
" " Dentistry	2,400 59	
" " Hygiene	391 17	
" " Veterinary Medicine	3,680 73	
" " " Hospital	7 37	
Bennett Hall	854 52	
Central Light and Heat Station	2,739 60	
		$84,537 87
Surplus:		
Department of Medicine	$4,909 39	
Department of Repairs	3,284 58	
Dormitories	14,533 34	
		22,727 31
Net deficit		$61,810 56

HOSPITAL.

GENERAL ACCOUNT.

RECEIPTS AND EXPENDITURES FROM AUGUST 31, 1899, TO AUGUST 31, 1900, INCLUSIVE.

Balance, August 31, 1899			$11,714 31
Receipts:			
Income from Invested Funds		$42,652 34	
Contributions for Endowment	$21,756 56		
Contributions not for Endowment	41,722 48		
		63,479 04	
Investments Paid Off or Sold		67,410 00	
Premium Received		40 00	
Receipts from Hospital		27,922 09	
Borrowed Money		11,855 47	
Interest on Bank Balances		301 64	
			213,660 58
Total Receipts			$225,374 89
Expenditures:			
Investments Made		$80,262 37	
Accrued Interest Paid		62 25	
Income and Funds not for Investment Paid to Hospital Superintendent		107,631 21	
Maintenance Pepper Clinical Laboratory		8,748 49	
Maintenance Maternity Hospital		796 00	
Buildings and Grounds		211 05	
James Dundas Clinic		6,098 67	
Loans Repaid		14,855 47	
Maennerchor Fund		853 00	
Taxes, Water Rent, Insurance, etc.		1,715 51	
Ladies' Fund Maternity Hospital		7 00	
Total Expenditures			221,241 02
Balance, August 31, 1900			$4,133 87
Of which there are:			
Investment Funds		$3,765 03	
Non-Investment Funds		368 84	
		$4,133 87	

HOSPITAL TRUST ACCOUNT.

Income from Trusts:		
General Fund (including Bed Funds)....	$14,764 46	
Hahn Ward Fund	3,168 86	
Alumni Ward Fund	480 00	
Mrs. G. W. Norris Fund	714 79	
Bement Fund	350 00	
Ward Chronic Diseases Fund (including Bed Funds)	6,998 16	
Pepper Clinical Laboratory Fund	9 00	
I. V. Williamson Hospital Fund..........	2,436 50	
Children's Ward Fund	1 00	
Henry Seybert Fund	32,21 05	
George S. Pepper Fund	2,647 19	
Recto-Genito-Urinary Fund	203 13	
Anna M. Powers Bed Fund	360 30	
Mary W. Eskens Bed Fund	300 00	
Dr. D. Hayes Agnew Hospital Fund	194 71	
Dr. D. Hayes Agnew Bequest for Care of Children	3,177 08	
Dr. D. Hayes Agnew Bequest for Maternity Hospital	60 00	
Andrew C. Butler Fund	240 00	
Ligonier Free Bed Fund	1,246 49	
Children's Orthopædic Ward Fund	760 00	
A. Sydney Roberts Apparatus Fund	160 00	
Ladies' Fund Maternity Hospital	7 00	
John White Field Bed Fund	215 40	
Abby White Peters Bed Fund	215 40	
Maternity Fund	430 82	
Edward Rhoads Memorial Fund	166 00	
Alfred Gustavus Baker Bed Fund	125 00	
		$42,652 34

DONATIONS TO THE UNIVERSITY.

ALUMNÆ ASSOCIATION FELLOWSHIP FOR WOMEN FUND.

Alumnæ Association of Women	$500 00	
		$500 00

AMERICAN SCHOOL AT ROME.

Cash	$250 00	
Harrison, C. C.	50 00	
Morgan, Randal	50 00	
Rosengarten, Joseph G.	50 00	
		400 00

ARCHÆOLOGY, DEPARTMENT OF.

Anonymous	$5,000 00	
		5,000 00

BOTANIC GARDEN FUND.

Botanical Society of Pennsylvania	$600 00	
Conard, Mrs. Mary B.	25 00	
Dreer, William F.	125 00	
Sims, John C.	25 00	
Trainor, John	42 92	
Welsh, William, Jr.	25 00	
		842 92

BOTANIC GARDEN FUND—ORCHID COLLECTION.

Brock, Robert C. H.	$48 00	
Manning, Katharine S.	10 00	
Through Professor J. M. Macfarlane	142 50	
		200 50

CHRISTIAN ASSOCIATION—UNIVERSITY OF PENNSYLVANIA.

Lewis, Lawrence	$500 00	
		500 00

DONATIONS TO THE UNIVERSITY.

DORMITORIES, NEW BUILDING FUND.

Bachmann, F. H.	$500 00	
Bailey, Banks & Biddle	250 00	
Baird, Thomas E.	1,000 00	
Barnes, William H.	500 00	
Bodine, Samuel T.	50,000 00	
Brock, Arthur	1,000 00	
Castner, S., Jr.	1,000 00	
Clyde, B. F.	250 00	
Ehret, Michael	1,000 00	
Frishmuth, E. H., Jr.	500 00	
Gerstley, William	100 00	
Godfrey, Lincoln	500 00	
Harrison, Mrs. Ellen Waln	50,000 00	
Hatfield, Walter	250 00	
Helme, William E.	250 00	
Hering, Walter E.	500 00	
Hinchman, Charles S.	1,000 00	
Jonas, Herman	100 00	
Kohn, Simon I. and Samuel	350 00	
Kurtz, W. W.	1,000 00	
Lesley, Robert W.	500 00	
Lindsay, James G.	100 00	
Lippincott, J. Dundas (Lippincott House)	2,500 00	
McFadden, J. Franklin	500 00	
Moore, E. J.	500 00	
Muhr, Jacob	500 00	
Presser, Theodore	200 00	
Shipley, Samuel R.	500 00	
Stewart, Mrs. Clara E.	250 00	
Townsend, Charles C.	500 00	
Wetherill, J. Price	500 00	
Wood, Alan, Jr.	1,000 00	
Wood, George	500 00	
Wood, R. Francis	300 00	
		$118,400 00

DORMITORIES—1887 HOUSE FUND.

Ashbrook, W. S.	$30 00
Benkert, W. C.	10 00
Clark, F. C.	20 00
Colladay, T. F.	5 00
Cowperthwaite, C. T.	50 00
Dolan, C. W.	80 00

DONATIONS TO THE UNIVERSITY.

Dunn, E. S.	$6 00	
Earle, Edgar P.	20 00	
Fetterolf, George	6 00	
Frazier, George H.	100 00	
Graham, F. W. W.	100 00	
Griscom, C. A., Jr.	25 00	
Hawkins, J. D.	12 00	
Huckel, Oliver	50 00	
Kenney, E. F.	15 00	
Latta, T. L.	30 00	
Magee, J. F., Jr.	15 00	
Martin, G. L.	5 00	
Pepper, David	2 50	
Pepper, G. W.	75 00	
Pryor, F. J.	5 00	
Smaltz, H. W.	60 00	
Smith, J. Somers, Jr.	25 00	
Townsend, C. C.	50 00	
Wagner, Joseph S.	10 00	
Weaver, John W.	2 50	
Whitaker, J. G. Neafie	50 00	
White, L. T.	10 00	
Wilson, Richard	10 00	
		$879 00

FELLOWSHIP IN ENGLISH FUND.

Sinnott, Joseph F.	$500 00	
		500 00

FUND FOR APPARATUS IN ELECTRO-CHEMISTRY.

Clark, E. W. & Co.	$1,000 00	
Morwitz, Joseph	100 00	
Rosengarten, Miss Fanny	250 00	
Welsh, Francis Ralston	500 00	
		1,850 00

GENERAL UNIVERSITY FUND.

Converse, John H.	$1,000 00	
Dignan, John	500 00	
Elkinton, Thomas	50 00	
Ely, Theodore N.	25 00	
Gimbel Brothers	500 00	
Greene, Stephen	2,500 00	
Harrison, Charles C.	50,000 00	

Donations to the University.

Harrison, William W.	$2,500 00	
Hope, James F.	250 00	
"J. S. H."	50 00	
Lit Brothers	500 00	
McCahan, W. J.	1,000 00	
Paul, Dr. James	100 00	
Paul, Miss Mary W.	300 00	
Pugh, Charles E.	500 00	
Rosengarten, Joseph G.	1,000 00	
Several friends of the University (annual for five years)	40,000 00	
Snellenburg, N.	500 00	
Spear, James	500 00	
Strawbridge & Clothier	1,000 00	
Thompson, S. S.	50 00	
Thouron, N.	280 00	
"W. H."	500 00	
		$103,605 00

GYMNASIUM—NEW BUILDING FUND.

"Anonymous"	$5,000 00	
Stotesbury, E. T.	500 00	
Thomas, George C.	500 00	
		6,000 00

JOHN STEWARDSON MEMORIAL FUND.

"A friend who desires to assure the Yearly award of the Scholarship"	$4,000 00	
"A friend to make up the deficiency in this Year's Scholarship"	86 18	
		4,086 18

LAW LIBRARY FUND.

Through Dr. William Draper Lewis	$15 00	
		15 00

LAW SCHOOL BUILDING FUND.

Addicks, William H.	$250 00	
Ballard, Ellis Ames	500 00	
Beck, James M.	250 00	
Biddle, Mrs. George W.	3,000 00	
Blum Brothers	500 00	
Brinton, Joseph Hill	250 00	
Burroughs, Joseph H.	125 00	

DONATIONS TO THE UNIVERSITY.

Cadwalader, John	$500 00
Cattell, Henry S.	250 00
Caven, Joseph L.	250 00
Clark, E. W.	250 00
Cope & Stewardson	120 00
Crawford & Loughlin	250 00
Dechert, Henry M.	500 00
Dickson, Samuel	2,500 00
Downing, R. W.	1,000 00
Frazier, George H.	500 00
Gerhard, John S.	250 00
Gest, John B.	1,000 00
Geyelin, H. Laussat	125 00
Gorman, William	250 00
Greene, Charles S.	100 00
Griffith, Warren G.	250 00
Griscom, C. A.	1,500 00
Harris, Joseph S.	1,000 00
Harrison, Mitchell	5,000 00
Harrity, William F.	250 00
Hipple, Frank K.	500 00
Jones, J. Levering	5,000 00
Junkin, Joseph de F., and H. S. P. Nichols	500 00
Lewis, F. D.	250 00
Lewis, Henry	250 00
Lex, William Henry	500 00
Lloyd, Malcolm	500 00
Lowrey, Dwight M.	125 00
McCaul, Charles	1,000 00
McKean, Mrs. Thomas	15,000 00
Magee, Horace	200 00
Massey, Walter S.	1,000 00
Moon, R. O.	500 00
Morgan, Charles E., Jr.	500 00
Morgan, Randal	2,500 00
Patterson, George Stuart	250 00
Paxson, Hon. E. M.	2,500 00
Pierce, George	250 00
Riley, Lewis A.	500 00
Rosengarten, Joseph G.	1,500 00
Smith, A. Lewis	500 00
Stewart, Ralph C.	250 00
Stickney & Conyngham	2,500 00
Strawbridge, Justus C.	2,500 00

DONATIONS TO THE UNIVERSITY.

Sullivan, James F.	$500 00	
Townsend, J. Barton	500 00	
Turner, William Jay	250 00	
"W. H."	1,000 00	
		$62,045 00

LAW SCHOOL BUILDING—OPENING CEREMONIES FUND.

Allinson, Edward P.	$25 00	
Beck, James M.	100 00	
Beeber, Dimner	100 00	
Biddle & Ward	50 00	
Bispham, George T.	25 00	
Bullitt, John C.	100 00	
Cadwalader, John	100 00	
Carson, Hampton L.	100 00	
Da Costa, Dr. J. M.	25 00	
Dale, Richard C.	100 00	
Dechert, Henry T.	50 00	
Dickson, Samuel	100 00	
Frazier, W. W.	100 00	
Freedley, A. T.	100 00	
Gest, John B.	100 00	
Gest, John M.	25 00	
Harris, Joseph S.	100 00	
Hart, Byerly	100 00	
Houston, Samuel F.	100 00	
Law Alumni	84 00	
McCrea, James	100 00	
Merrick, J. Vaughan	100 00	
Morgan & Lewis	100 00	
Morgan, Randal	100 00	
Price, Eli Kirk	20 00	
Read & Pettit	100 00	
Rosengarten, Joseph G.	100 00	
Rothermel, P. F.	100 00	
Scott, William	100 00	
Sellers, William	100 00	
Simpson & Brown	100 00	
Smith, Walter George	100 00	
Staake, William H.	100 00	
Through John C. Bell	110 59	
Todd, M. Hampton	100 00	
		3,014 59

DONATIONS TO THE UNIVERSITY.

LIBRARY—CURRENT EXPENSE FUND.

Carson, Hampton L.	$300 00	
Jeanes, Joseph Y.	250 00	
Potter, William	28 36	
		$578 36

LIBRARY—DRAMATIC FUND.

Graduate Department in English	$17 00	
		17 00

LIBRARY—ENGLISH PARLIAMENTARY REPORTS FUND.

Gibson, Miss Mary K.	$1,000 00	
		1,000 00

LIBRARY—MUSIC FUND.

Presser, Theodore	$25 00	
		25 00

LIBRARY—PATHOLOGY FUND.

Beates, Dr. Henry, Jr.	$25 00	
		25 00

LIBRARY—WARREN FUND.

Warren, E. Burgess	$1,000 00	
		1,000 00

LOCUST STREET IMPROVEMENT FUND.

Busch, Henry Paul	$10 00	
Busch, Miers	10 00	
"Cash"	200 00	
Carson, Hampton L.	250 00	
Gilpin, Bernard	10 00	
Jeanes, Isaac W.	250 00	
McElroy, Clayton	50 00	
Reed, Alan H.	10 00	
Rosengarten, Joseph G.	100 00	
Van der Wielen, Adolph	100 00	
		990 00

MECHANICAL AND ELECTRICAL ENGINEERING SCHOOL IMPROVEMENT FUND.

Merrick, J. Vaughan	$1,567 80	
		1,567 80

DONATIONS TO THE UNIVERSITY.

MECHANICAL AND ELECTRICAL ENGINEERING DEPARTMENT ADVERTISING FUND.

Wood, R. D. & Co.	$300 00	
		$300 00

MEMORIAL TOWER FUND.

"A friend to the University"	$15,000 00
Barnes, J. Sellers	5 00
Bartlett, Edward T.	5 00
Bartram, F. de M.	5 00
Beeber, William P.	1 00
Bell, John C.	25 00
Brinton, Clarence C.	2 00
Chambers, George H.	5 00
Chipley, C. E.	5 00
Clark, A. W.	10 00
Cohen, Henry B.	5 00
Crawhall, G. W.	2 00
Currie, Thomas R.	2 00
Curtin, Roland T.	5 00
Dechert, Henry T.	25 00
Dick, Lewis R.	25 00
Dickson, Hazard	10 00
Donges, John S.	10 00
Durham, John S.	50 00
Egle, William H.	5 00
Fisher, F. W.	2 00
Forsyth, James Perkins	5 00
Foulke, Roland R.	5 00
Fraley, Frederick, Jr.	5 00
Fraley, Joseph C.	10 00
Gabb, Isaac C.	5 00
Gillespie, J. L.	2 00
Girvin, Dr. John H.	10 00
Graham, Robert	2 00
Groome, D. M.	5 00
Gucker, Frank T.	20 00
Guggenheim, William	100 00
Harrison, Charles C.	1,000 00
Hirst, Dr. Barton C.	25 00
Hornor, William Macpherson	100 00
Humphreys, Richard L.	5 00
James, Fleming	5 00
Jastrow, Morris, Jr.	15 00
Jones, George M.	10 00
Koch, Theodore W.	10 00

DONATIONS TO THE UNIVERSITY.

Kuemmerle, G. C.	$2 00	
Lawrence, Frank W.	3 00	
Louisson, William S.	5 00	
McCollin, E. G.	50 00	
McDonald, Silas, Jr.	10 00	
McFadden, J. J.	10 00	
McKeehan, Charles L.	100 00	
Moore, H. W.	25 00	
Morgan, J. B.	50 00	
Neilson, Lewis	10 00	
Newlin, Arthur	5 00	
Paddock, Ralph L.	10 00	
Page, Henry	10 00	
Patterson, Joseph H.	5 00	
Pierce, Harold	50 00	
Potter, Thomas C.	100 00	
Potts, H. Cameron	5 00	
Redmond, Dr. H.	5 00	
Remington, J. Percy	5 00	
Reynolds, F. P.	5 00	
Rosengarten, Adolph G.	100 00	
Rosengarten, Miss Fanny	100 00	
Rosengarten, Frank H.	100 00	
Rosengarten, George D.	100 00	
Rosengarten, Joseph G.	500 00	
Rosengarten, Joseph G., Jr.	100 00	
Russell, Randolph P.	10 00	
Ryan, Charles J.	9 00	
Saunders, W. L.	100 00	
Schmid, H. Ernst	10 00	
Sims, John C.	25 00	
Smith, Benjamin R.	5 00	
Smith, Walter Bassett	5 00	
Snyder, George B.	5 00	
Stewart, R. C.	10 00	
Stewart, T. S.	10 00	
Stout, George C.	5 00	
Strickland, D. Hayes	10 00	
Teck, Edward H.	20 00	
Transom, Fred.	5 00	
Vail, Louis de Pui	5 00	
Van der Weilen, Adolph B.	20 00	
Voight, A. G.	5 00	
White, W. W.	25 00	
Wolf, Otto C.	50 00	
		$18,342 00

Donations to the University.

NATHANIEL CHAPMAN MEMORIAL.

Estate of Samuel Jackson, deceased	$2,850 00	
		$2,850 00

NEW YORK ALUMNI HOUSE.

Guggenheim, William	$2,500 00	
Towne, Henry R.	1,000 00	
Ward, Henry Galbraith	1,000 00	
		4,500 00

NEEDY STUDENTS' FUND.

Collins, Henry H.	$1,000 00	
Fries, Aaron	300 00	
Harrison, Mrs. Letitia H.	100 00	
Ladies' Home Journal	1,000 00	
		2,400 00

NEW MEDICAL LABORATORY FUND.

Castner, S., Jr.	$1,000 00	
Clothier, Isaac H.	500 00	
Ivins, William	500 00	
Kneass, Dr. S. S.	125 00	
Small, Dr. J. Frank	75 00	
Stengel, Dr. Alfred	500 00	
Stewart, Thomas S.	250 00	
Tyson, Dr. James	125 00	
Thomson, Dr. William	500 00	
White, Dr. J. William	1,000 00	
Yarrow, Dr. Thomas J.	500 00	
Zimmerman, Dr. M. W.	125 00	
		5,200 00

PEPPER HALL ENDOWMENT FUND.

Pepper, Mrs. Frances S.	$50,000 00	
		50,000 00

PRIZE FUNDS.

Comegys Prize—B. B. Comegys	$50 00	
Delano Prize—Eugene Delano	50 00	
Latin Prize—Alumni Society, College	22 00	
		122 00

PROFESSORS' RETIRING FUND.

Da Costa, Dr. J. M.	$1,000 00	
Drexel, Mrs. Lucy W.	1,000 00	
		2,000 00

DONATIONS TO THE UNIVERSITY.

PHYSICAL LABORATORY FUND.

"A friend to Science"	$25,000 00	
		$25,000 00

SAMUEL JACKSON MEMORIAL FUND.

Estate of Margaret M. Jackson, deceased	$2,880 00	
		2,880 00

SCHOLARSHIPS—GENERAL FUND.

"Anonymous"	$10,000 00	
Harris, Mrs. Mary Powers	2,500 00	
Powers, Thomas Harris	1,000 00	
Roberts, the Misses E. C. and Frances A.	2,000 00	
		15,500 00

SCHOLARSHIP—DEPARTMENT OF BIOLOGY.

"J. S. H."	$240 00	
		240 00

SCHOLARSHIP—THE JAMES LOGAN

"Anonymous"	$5,000 00	
		5,000 00

SCHOLARSHIP—THE LOUISE HARRISON

Harrison, Thomas S.	$5,000 00	
		5,000 00

SCHOLARSHIP—THE ROBERT MORRIS

"Anonymous"	$5,000 00	
		5,000 00

SEMITIC DEPARTMENT.

Sulzberger, Hon. Mayer	$100 00	
		100 00

UNIVERSITY FUND FOR WOMEN.

"J. S. H."	$100 00	
		100 00

UNIVERSITY PUBLICATION FUND.

Alumni Society, College	$100 00	
Cohen, Charles J.	100 00	
Craig, John F.	50 00	

DONATIONS TO THE UNIVERSITY.

Cuyler, T. De Witt	$100 00	
Dale, Richard C.	100 00	
Dickson, Samuel	100 00	
Dingee, John H.	100 00	
Duane, James May	100 00	
Erringer, J. L.	100 00	
Gowen, Francis I.	100 00	
Jones, Thomas F.	100 00	
Joyce, William H.	100 00	
McFadden, George	50 00	
McFadden, J. Franklin	100 00	
Riebenack, M.	100 00	
Rosengarten, Joseph G.	200 00	
Sims, John C.	100 00	
Sparhawk, John, Jr.	100 00	
Stilz, John	100 00	
Vaux, J. Waln	200 00	
		$2,100 00

VETERINARY HOSPITAL FUND.

Goodwin, Mrs. Josephine S.	$1,000 00	
Lippincott, J. Bertram	1,000 00	
Lippincott, Walter	500 00	
		2,500 00

VIVARIUM FUND.

Page, Harlan	$5,000 00	
		5,000 00
		$467,175 35

DONATIONS TO THE HOSPITAL.

ALFRED GUSTAVUS BAKER BED FUND.

Baker, Dr. George Fales	$5,000 00	
		$5,000 00

ABRAHAM AND MARY BAKER BED FUND.

Estate of Elizabeth Wirt Baker, deceased	$5,000 00	
		5,000 00

MOULTON H. FORREST BED FUND.

Mrs. Emma L. Forrest	$600 00	
		600 00

CHILDREN'S MEDICAL WARD FUND.

Through Mrs. K. E. Langstroth	$740 00	
		740 00

CHILDREN'S ORTHOPÆDIC WARD FUND.

Austin, Mrs. James S.	$200 00	
Clark, Mrs. E. W.	200 00	
Collins, Mrs. H. H.	200 00	
Conrad, Mrs. T. K.	200 00	
Dickson, Mrs. Samuel	200 00	
Freedley, Mrs. J. K.	200 00	
Freedley, Mrs. William G.	200 00	
Harris, Mrs. Mary Powers	500 00	
Porter, Mrs. W. W.	200 00	
Reed, Mrs. Anna L.	200 00	
Saturday Club of New Jersey	200 00	
"The Fairy Bed"	100 00	
Townsend, Mrs. John W.	200 00	
		2,800 00

GENERAL FUND.

Allison Manufacturing Company	$150 00	
Charity Ball Committee	3,000 00	
City of Philadelphia	375 00	
State of Pennsylvania	22,344 69	
Sundry donations at Hospital	392 20	
Welsh, Mrs. J. Lowber	50 00	
		26,311 89

DONATIONS TO THE HOSPITAL.

PEPPER CLINICAL LABORATORY FUND.

Estate of William Pepper, M. D., deceased....	$30 00	
		$30 00

PEPPER CLINICAL LABORATORY MAINTENANCE FUND.

Through Dr. Alfred Stengel	$8,820 59	
		8,820 59

PRIVATE FLOOR FUND—THROUGH MRS. C. C. HARRISON.

Clark, Clarence H.........................	$100 00	
Dolan, Thomas	1,000 00	
Drexel, George W. C.......................	200 00	
Harrison, Mrs. C. C.......................	250 00	
Paul, James W., Jr........................	100 00	
Paul, Miss Mary W.........................	100 00	
Van Rensselaer, Mrs. Alexander	250 00	
		2,000 00

ROOF GARDEN FUND—THROUGH MRS. C. C. HARRISON.

Harris, Mrs. Mary Powers	$300 00	
Welsh, Miss Fanny de L....................	100 00	
		$400 00

STABLE REPAIR FUND.

Janney, E. K..............................	$50 00	
		50 00

CHRONIC DISEASES WARD FUND.

Estate of Margaret Justin, deceased	$11,726 56	
		11,726 56
		$63,479 04

The following contributions were received just after closing the books for the fiscal year, being donations to the Memorial Gate Fund, Class of '94, through William G. Moore, Treasurer of the fund.

Alden, R. M..............................	$10 00
Arnold, J. K.............................	5 00
Belknap, Henry H.........................	10 00

Donations, Class of '94.

Blackburne, Clarence E.	$5 00
Brinton, Frederick S.	5 00
Brines, William P.	1 00
Brown, Francis H.	25 00
Bullock, J. Maxwell	50 00
Cauffman, William L.	20 00
Coates, George M.	20 00
Codman, George D.	20 00
Collins, Henry H., Jr.	20 00
Coombs, Robert D., Jr.	5 00
Coxe, E. B., Jr.	25 00
Craig, Frank A.	5 00
Denckla, Herman A.	20 00
Du Pont, Francis I.	15 00
Fetterolf, Edwin H.	20 00
Francine, A. P.	100 00
Fuguet, Howard	25 00
Gates, Elias	5 00
Gittings, J. G.	10 00
Heysham, Theodore	5 00
Herrick, Cheesman A.	20 00
Hirsh, Walter A.	5 00
Innes, R. H.	20 00
Jefferys, William Hamilton	3 00
Kindelberger, E. Crosby	20 00
MacLain, Christian S.	25 00
Moore, William G.	100 00
Murtagh, J. C.	10 00
Paddock, Ernest M.	2 00
Partridge, Charles L.	5 00
Pepper, William	100 00
Phillips, James M.	5 00
Quinn, Arthur H.	10 00
Riebenack, H. G.	10 00
Rosengarten, J. Clifford	50 00
Rowland, Benjamin	25 00
Stackhouse, Powell	5 00
Sherlock, A. E.	5 00
Spear, Raymond	10 00
Sinclair, Robert S.	20 00
Smith, Thomas	5 00
Smith, Thomas Kilby	20 00
Thomas, George C.	25 00
Thompson, Walter S.	10 00

DONATIONS, CLASS OF '94.

Tull, Samuel P.	$25 00	
Van Lennep, Augustus	15 00	
Welch, Charles W.	10 00	
Wiedershiem, Theodore E., Jr.	50 00	
Wilson, Percy N.	50 00	
Wood, Alan, 3d.	2 00	
Woodman, Henry G.	10 00	
Interest on deposits	28 57	
		$1,126 57

GEORGE L. HARRISON, JR., *Treasurer.*

INDEX.

Made in United States
Troutdale, OR
11/03/2026

56844531R00146